Optical
mineralogy

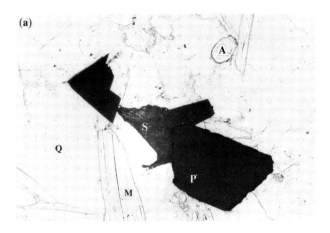

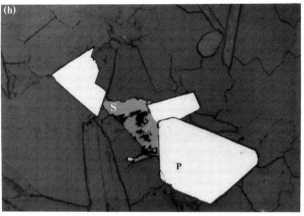

Frontispiece Photomicrographs, taken using (a) transmitted light and (b) reflected light, of the same area of a polished thin section of quartzite containing pyrite (P), sphalerite (S), muscovite (M), apatite (A) and abundant quartz (Q).

The features illustrated in transmitted light are: (i) opacity – pyrite is the only opaque phase, sphalerite is semi-opaque, and the others are transparent; (ii) relief – very high (sphalerite, $n \approx 2.4$), moderate (apatite, $n \approx 1.65$), moderate (muscovite, $n \approx 1.60$), and low (quartz, $n \approx 1.55$); (iii) cleavage – perfect in muscovite (n is the refractive index of the mineral).

The following values of reflectance (the percentage of incident light reflected by the mineral) are illustrated in reflected light: pyrite, 54% (white – true colour slightly yellowish white); sphalerite, 17% (grey); apatite, 6% (dark grey); and muscovite and quartz, both 5% (dark grey).

Note that opaque grains, grain boundaries and cleavage traces appear black in transmitted light, whereas pits (holes), grain boundaries and cleavage traces appear black in reflected light.

Optical mineralogy

Principles and practice

C. D. Gribble
A. J. Hall

Department of Geology & Applied Geology,
University of Glasgow

UCL PRESS
UCL
PRESS
Taylor & Francis Group

First published in 1992 by UCL Press.
Reprinted 1993, 1997, 1999, 2001

UCL Press Limited
11 New Fetter Lane
London EC4P 4EE

UCL Press is an imprint of the Taylor & Francis Group

The name of University College London (UCL) is a registered
trade mark used by UCL Press with the consent of the owner.

ISBNS: 1-85728-013-X HB, 1-85728-014-8 PB

A CIP catalogue record for this book is available from the British Library.

Front cover illustration
A thin section of dolerite under crossed polars in a transmitted-light microscope. The
small black & white crystals of plagioclase feldspar are enclosed by large coloured
crystals of augite. (Courtesy of the Department of Geology & Applied Geology,
University of Glasgow)

Typeset in Times Roman by Computate (Pickering) Ltd, North Yorkshire
Printed and bound by Biddles Ltd, King's Lynn and Guildford, England

Contents

2 Silicate minerals

3 The non-silicates

4 Transmitted-light crystallography

5   Reflected-light theory

Preface

This book is the successor to *A practical introduction to optical mineralogy*, which was written in the early 1980s, and published by George Allen & Unwin in 1985.

Our intention, once again, is to introduce the student of geology to the microscopic examination of minerals, by both transmitted and reflected light. These techniques should be mastered by students early in their careers, and this text has been proposed in the full awareness that it will be used as a laboratory handbook, serving as a quick reference to the properties of minerals. However, care has been taken to present a systematic explanation of the use of the microscope, as well as to include an extended explanation of the theoretical aspects of optical crystallography in transmitted light. The book is therefore intended as a serious text that introduces the study of minerals under the microscope to the intending honours student of geology, as well as providing information for the novice or interested layman.

Both transmitted-light and reflected-light microscopy are considered, the former involving examination of transparent minerals in thin section and the latter involving examination of opaque minerals in polished section. Reflected-light microscopy is of increasing importance in undergraduate courses dealing with ore mineralization, but the main reason for combining both aspects of microscopy in this text is that they are both important in petrographic studies of rocks, especially where opaque minerals constitute a significant part of a rock's volume. Dual-purpose microscopes are readily available and are ideal for the study of polished thin sections, which is a necessary prerequisite of electron microprobe analysis, in which a narrow electron beam, about 1 μm in diameter, is used to analyze a point on the polished surface of a mineral. The sections on reflected-light microscopy have also been written to assist any experienced transmitted-light microscopists who may wish to gain a better understanding of these techniques.

Chapter 1 provides an introduction to the study of minerals under the microscope, using both transmitted- and reflected-light techniques. Chapter 2 deals with detailed descriptions of the silicate minerals in transmitted light, and includes all of the minerals likely to be encountered in thin sections of most normal rocks. Chapter 3 includes occasional descriptions of some non-silicates in transmitted light, but mostly deals with reflected-light descriptions of the opaque non-silicate minerals. In both Chapters 2 and 3 the minerals are presented in broadly alphabetical order, but closely related minerals are grouped together: the best way to locate a particular mineral is to make use of the index at the end of the book. Chapter 4 gives a detailed account of transmitted-light optical crystallography, since it is felt – by ourselves and many others – that students should have access to detailed explanations of the various optical properties. In a similar manner, Chapter 5 gives an account of reflected-light theory. Both of these theoretical chapters have deliberately been placed after the main mineral descriptions of Chapters 2 and 3. The appendices include systematic lists of the optical properties of minerals, for use in identification.

As this book is intended as an aid to the identification of minerals in rock thin sections, by either transmitted-light or reflected-light techniques, the description or interpretation of mineral relationships is not discussed, although the mineral descriptions in Chapters 2 and 3 contain brief notes on occurrence.

We hope that our text fulfils its purpose, and that all students of geology find it helpful and enjoyable to use. Any comments with regard to improvements in future editions would be most welcome.

C. D. Gribble & A. J. Hall
University of Glasgow

Acknowledgements

As in our previous book, the sections on transmitted light have been written by C. D. Gribble, who again acknowledges the debt he owes to Kerr (1977), whose format has been employed in the descriptions in Chapter 2, and to many other authors whose information has been used in compiling the optical data used in the text (particularly Deer, Howie & Zussman 1966, 1978 *et seq*, Bloss 1971, Phillips & Griffin 1981, Smith 1974, and Wahlstrom 1959).

The descriptions of the opaque minerals, by A. J. Hall, are based upon data in many texts, particularly the tables of Uytenbogaardt & Burke (1971), the classic text *Dana's system of mineralogy*, edited by Palache et al. (1962), the excellent descriptions of ore minerals by Ramdohr (1969), and the atlas of Picot & Johan (1982). The textbook on the microscopic study of minerals by Galopin & Henry (1972) and various publications by Cervelle form the basis of Chapter 5.

We are grateful for the support of our colleagues at the University of Glasgow while writing the text, and comments and suggestions by B. E. Leake proved most helpful. We particularly appreciated the kindly shepherding of our text by Roger Jones, as UCL Press was in the process of being formed, and wish to thank all those concerned in the final editing of the text: of course, any inaccuracies and errors are our own. We wish to thank Chapman & Hall for permission to use illustrations from *A Practical Introduction to Optical Mineralogy*.

List of tables

List of plates

Symbols and abbreviations used in the text

Crystallographic properties of minerals

$a\,b\,c$ or $X\,Y\,Z$	crystallographic axes
hkl	Miller's indices, which refer to crystallographic orientation
(111)	a single plane or face
{111}	a form; all planes with same geometric relationship to axes
[111]	zone axis; planes parallel to axis belong to zone
β	angle between a and c in the monoclinic system
α, β, γ	angles between b and c, a and c, and a and b in the triclinic system

Light

λ	wavelength
A	amplitude
PPL	plane or linearly polarized light

Microscopy

N, S, E, W	north (up), south (down), east (right), west (left) in image or in relation to crosswires
NA	numerical aperture
XPOLS, XP, CP	crossed polars (analyser inserted)

Optical properties

n or RI	refractive index of mineral
N	refractive index of immersion medium
n_o	RI of ordinary ray
n_e	RI of extraordinary ray
n_α	minor RI
n_β	intermediate RI
n_γ	major RI
o	ordinary ray vibration direction of uniaxial mineral
e	extraordinary ray vibration direction of uniaxial mineral
α, β, γ	principal vibration directions of general optical indicatrix
δ	maximum birefringence ($n_e - n_o$ or $n_\gamma - n_\alpha$)
$2V$	optic axial angle
$2V_\alpha$	optic axial angle bisected by α
$2V_\gamma$	optic axial angle bisected by γ
Bx_a	acute bisectrix (an acute optic axial angle)
Bx_o	obtuse bisectrix (an obtuse optic axial angle)
OAP	optic axial plane
$\gamma\hat{\ }cl$	angle between γ (slow component) and cleavage
$\alpha\hat{\ }cl$	angle between α (fast component) and cleavage

k	absorption coefficient
R	reflectance (usually expressed as a percentage, $R\%$)
R_{min}	minimum reflectance of a polished section of a bireflecting mineral grain
R_{max}	maximum reflectance of a polished section of a bireflecting mineral grain
R_o	principal reflectance corresponding to ordinary ray vibration direction of a uniaxial mineral
R_c	principal reflectance corresponding to extraordinary ray vibration direction of a uniaxial mineral
ΔR	bireflectance ($R_{max} - R_{min}$) referring to individual section or maximum for mineral

Quantitative colour

$Y\%$	visual brightness
λ_d	dominant wavelength
$P_e\%$	saturation
x, y	chromaticity co-ordinates

Mineral properties

VHN	Vickers hardness number
H	hardness on Mohs' scale
D	density
SG	specific gravity

General

P	pressure
T	temperature
XRD	X-ray diffraction
REE	rare earth elements
nm	nanometre
μm	micrometre or micron
mm	millimetre
cm	centimetre
d	distance or length
Å	angstroms
cl	cleavage
kb	kilobar
$>$	greater than
$<$	less than
$\geq$	greater than or equal to
$\leq$	less than or equal to
$\sim$	approximately
$\approx$	approximately equal to
$\perp$	perpendicular to
$\parallel$	parallel to
$4 +$	four or greater
3D	three dimensional
Zn + Fe + S	association of elements in ternary chemical system
Zn–Fe–S	association of elements

XV

1 Introduction to the microscopic study of minerals

1.1 Introduction

Microscopes vary in their design, not only in their appearance but also in the positioning and operation of the various essential components. These components are present in all microscopes and are described briefly below. Although dual-purpose microscopes incorporating both transmitted- and reflected-light options are now available (Fig. 1.1), it is more convenient to describe the two techniques separately. More details on the design and nature of the components can be obtained in textbooks on microscope optics.

1.2 The transmitted-light microscope

The light source   In transmitted-light studies a lamp is commonly built into the microscope base (Fig. 1.2). The typical bulb used has a tungsten filament (A source) which gives the field of view a yellowish tint. A blue filter can be inserted immediately above the light source to change the light colour to that of daylight (C source).

In older microscopes the light source is quite separate from the microscope and is usually contained in a hooded metal box, to which can be added a blue glass screen for daylight coloured light. A small movable circular mirror, one side of which is flat and the other concave, is attached to the base of the microscope barrel. The mirror is used to direct the light through the rock thin section on the microscope stage, and the flat side of the mirror should be used when a condenser is present.

The polarizer   The assumption is made that light consists of electromagnetic vibrations. These vibrations move outwards in every direction from a point source of 'white' light, such as a microscope light. A polarizing film (the polarizer) is held within a

1

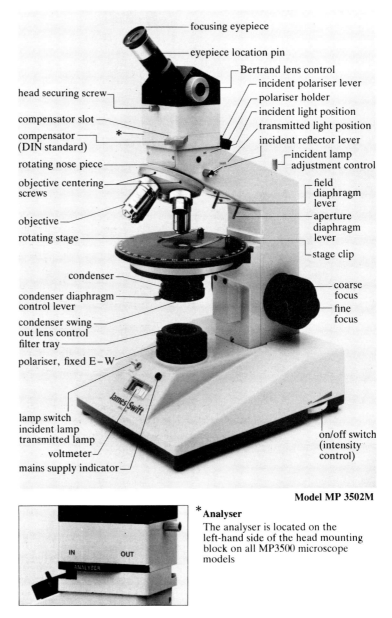

Figure 1.1   The Swift Student polarizing microscope (photo courtesy of Swift Ltd).

lens system located below the stage of the microscope, and this is usually inserted into the optical path. On passing through the polarizer the light is 'polarized' and now vibrates in a single plane. This is called *plane polarized light* (PPL). In most UK microscopes

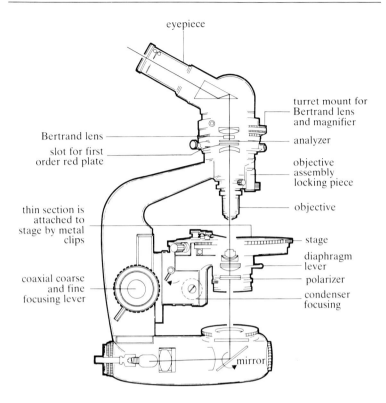

Figure 1.2 A modern transmitted-light microscope. Older models may focus by moving the upper barrel of the microscope (not the stage, as in the illustration), and may use an external light source. The illustration is based on a Nikon model POH-2 polarizing microscope.

the polarizer is oriented to give east–west-vibrating incident light (see also Ch. 4).

Substage diaphragms One or two diaphragms may be located below the stage. The field diaphragm, often omitted on simple student microscopes, is used to reduce the area of light entering the thin section, and should be in focus at the same position as the thin section: it should be opened until it just disappears from view. The aperture diaphragm is closed to increase resolution: it can be seen when the Bertrand lens is inserted.

The condenser or convergent lens A small circular lens (the condenser) is attached to a swivel bar, so that it can be inserted into the optical train when required. It serves to direct a cone of light onto the thin section and give optimum resolution for the objectives used.

The entire lens system below the microscope stage, including the polarizer, aperture diaphragm and condenser, can often be racked upwards or downwards in order to optimize the quality of illumination. However, some microscopes do not possess a separate convergent lens and, when a convergent lens is needed, the substage lens system is racked upwards until it is just below the surface of the microscope stage.

Stage The microscope stage is flat and can be rotated. It is marked in degrees, and a side vernier enables angles of rotation to be measured accurately. The stage can usually be locked in place at any position. The rock thin section is attached to the centre of the stage by metal spring clips.

Objectives Objectives are magnifying lenses with the power of magnification inscribed on each lens (e.g. × 5, × 30). An objective of very high power (e.g. × 100) usually requires an immersion oil between the objective lens and the thin section.

Eyepiece The eyepiece (or ocular) contains crosswires which can be independently focused by rotating its uppermost lens. Eyepieces of different magnification are available. Monocular heads are standard on student microscopes. Binocular heads may be used and, if correctly adjusted, reduce eye fatigue.

The analyser The analyser is similar to the polarizer: it is also made of polarizing film but oriented in a north–south direction, i.e. at right angles to the polarizer. When the analyser is inserted into the optical train, it receives light vibrating in an east–west direction from the polarizer, which it cannot transmit; thus the field of view is dark and the microscope is said to have *crossed polars* (CP, XPOLS or XP). With the analyser out, only the polarizer is in position; plane polarized light is being used and the field of view appears bright.

The Bertrand lens The Bertrand lens is used to examine interference figures (see Section 1.4.2). When it is inserted into the upper microscope tube an interference figure can be produced which fills the field of view, provided that the convergent lens is also inserted into the optical path train.

The accessory slot Below the analyser is a slot into which accessory plates, e.g. quartz wedge, or first-order red, can be inserted. The slot is oriented so that accessory plates are inserted at 45° to the crosswires. In some microscopes it may be possible to rotate the slot.

Focusing The microscope is focused by moving the microscope stage either up or down (newer models) or by moving the upper microscope tube up or down (older models). Both coarse and fine adjusting knobs are present.

1.3 The appearance of thin sections using transmitted-light microscopy

Thin sections should be placed on the stage with the cover slip up, and observation should commence with *low magnification* and *plane polarized light* (analyser out). The section should be focused with the aperture diaphragm opened up only slightly, so that the image does not appear too 'flat'. The field should appear generally white, but it will exhibit grain colour and textures depending on the nature of the rock being examined.

The following features may be observed (see Plates 1a, 2a and 4a):

(a) Transparent phases (minerals, glass or liquid) appear white to coloured, because they transmit most of the source light. More details of the observations that can be made on transparent minerals are given in Section 1.4.

(b) Absorbing phases (opaques or ore minerals) appear black, because the light is absorbed. Some minerals may transmit a small amount of light; this is often coloured red to brown because longer wavelengths of light are often absorbed to a lesser extent than light at the blue end of the spectrum.

(c) Grain boundaries, cleavage traces and microfractures appear as thin black lines, because light is scattered and refracted at such boundaries.

(d) Fluid inclusions (Shepherd et al. 1985) can sometimes be observed within minerals. These tend to be small irregular rounded areas, often as groups or in zones. Careful observation with high magnification sometimes reveals small gas bubbles which move in an erratic manner.

(e) Holes, fractures and patches where the rock section is missing appear white; the light is simply transmitted through the glass and mounting medium. Such areas are usually 'isotropic' but sometimes appear to be weakly anisotropic, depending on the type of cement used.

(f) Artefacts which may be observed include the following. (i) Bubbles in the mounting medium: these can be small circles within cavities, or in fractures filled with the cementing medium, or they may appear superimposed on the minerals within the section or as large amoeboid areas, sometimes with

5

an enhanced 'relief'. Such areas may 'carry' the optical properties of the mineral(s), which can be above or below the bubble. (ii) Preparation materials, such as grinding grit, usually appear as black dust or transparent to dark blue-black angular grains of high relief, which tend to accumulate in fractures or at the edges of the section. Isolated fragments of corundum can sometimes appear to be inherent to the thin section, but they are usually out of context in terms of texture and association.

1.4 Systematic description of minerals in thin section using transmitted light

Descriptions of transparent minerals are given in a systematic manner in Chapters 2 and 3, and the terms used are explained below. The optical properties of each mineral include some determined in plane polarized light and others determined using crossed polars. For most properties a low-power objective (up to × 10) is used.

1.4.1 Properties in plane polarized light

The analyser is taken out of the optical path to give a bright image (see Frontispiece).

Colour   Minerals show a wide range of colour (by which we mean the natural or 'body' colour of a mineral), from colourless minerals (such as quartz and feldspars) to coloured minerals (such as brown biotite, yellow staurolite and green hornblende). Colour is related to the wavelength of visible light, which ranges from violet (wavelength, $\lambda = 0.00039$ mm or 390 nm) to red ($\lambda = 760$ nm). White light consists of all the wavelengths between these two extremes. In thin section, white light passes unaffected through colourless minerals (such as quartz) and none of its wavelengths is absorbed, whereas with opaque minerals (such as metallic ores) all wavelengths are absorbed and the mineral appears black. With coloured minerals, selective absorption of wavelengths takes place and the colour seen represents a combination of the wavelengths of the light transmitted by the mineral.

Pleochroism   When the microscope stage is rotated, some coloured minerals change colour between two 'extremes', each of which is seen twice during a complete (360°) rotation. Such a mineral is said to be pleochroic, and ferromagnesian minerals, such as the amphiboles, biotite and staurolite of the common rock-forming silicates, possess this property.

Pleochroism is explained by the unequal absorption of light by the mineral in different orientations. For example, in a longitudinal section of biotite, when plane polarized light from the polarizer enters the mineral which has its cleavages parallel to the vibration direction of the light, considerable absorption of light occurs and the biotite appears dark brown. If the mineral section is then rotated through 90° so that the plane polarized light from the polarizer enters the mineral with its cleavages now at right angles to the vibration direction, much less absorption of light occurs and the biotite appears pale yellow (see Plate 1a).

Habit This refers to the shape that a particular mineral exhibits in different rock types. A mineral may appear euhedral, with well defined crystal faces, or anhedral, where the crystal has no crystal faces present, such as when it crystallizes into gaps left between crystals that formed earlier. Other descriptive terms include prismatic, when the crystal is elongate in one direction, acicular, when the crystal is needle-like, or fibrous, when the crystals resemble fibres. Flat, thin crystals are termed tabular or platy.

Cleavage Most minerals can be cleaved along certain specific crystallographic directions which are related to planes of weakness in the mineral's atomic structure. These planes or cleavages, which are straight, parallel and evenly spaced in the mineral, are denoted by Miller's indices, which indicate their crystallographic orientation. Some minerals such as quartz and garnet possess no cleavages, whereas others may have one, two, three or four cleavages. When a cleavage is poorly developed it is called a parting. Partings are usually straight and parallel but *not* evenly spaced. The number of cleavages seen depends upon the orientation of the mineral section. Thus, for example, a prismatic mineral with a square cross section may have two prismatic cleavages. These cleavages are seen to intersect in a mineral section cut at right angles to the prism zone, but in a section cut parallel to the prism zone the traces of the two cleavages are parallel to each other and the mineral appears to possess only one cleavage (e.g. pyroxenes and andalusite).

Relief All rock thin sections are trapped between two thin layers of resin (or cementing material) to which the glass slide and the cover slip are attached. The refractive index (RI) of the resin is 1.54. The surface relief of a mineral is essentially constant (except for carbonate minerals), and depends on the difference between the RI of the mineral and the RI of the enclosing resin: the greater the difference between the RI of the mineral and the resin, the rougher the appearance of the surface of the mineral. This is because the

7

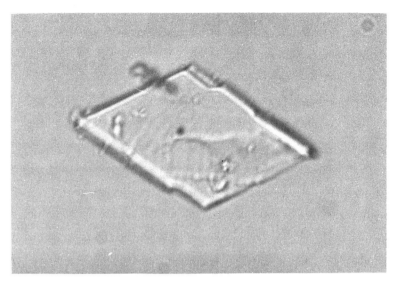

(a)

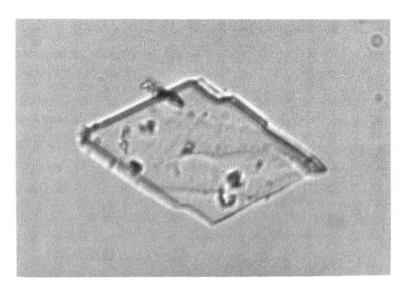

(b)

Figure 1.3(a) A transmitted-light photomicrograph (PPL) of a crystal showing the Becke line (thin white line). The microscope stage has been lowered slightly (equivalent to racking the microscope barrel upwards) and the white line has moved into the crystal, confirming that $RI_{crystal} > RI_{surrounding\ oil}$. This conclusion is valid only for this orientation of the crystal if it is not of cubic symmetry. **(b)** As (a), but the microscope stage has been raised slightly and the white line has moved into the oil surrounding the crystal, again confirming that $RI_{crystal} > RI_{surrounding\ oil}$.

surfaces of the mineral in thin section are made up of tiny elevations and depressions which reflect and refract the light.

If the RIs of the mineral and resin are similar the surface appears to be smooth. Thus, for example, the surfaces of garnet and olivine, which have much higher RIs than the resin, appear rough whereas the surface of quartz, which has the same RI as the resin (1.54), is smooth and virtually impossible to detect. Low, medium and high relief is demonstrated in Plate 1a.

To obtain a more accurate estimate of the RI of a mineral (compared to 1.54) a mineral grain should be found at the edge of the thin section, with its edge against the cement. The diaphragm of the microscope should be adjusted until the edge of the mineral is clearly defined by a thin, bright band of light which is exactly parallel to the mineral boundary. The microscope tube is then carefully racked upwards (or the stage lowered), and this thin band of light – the Becke line – will appear to move towards the medium with the higher RI. For example, if $RI_{mineral}$ is greater than RI_{cement} the Becke line will appear to move into the mineral when the microscope tube is slowly racked upwards (see Figs 1.3a & b). If the RI of a mineral is close to that of the cement, then the mineral surface will appear smooth, and dispersion of the refractive index may result in slightly coloured Becke lines appearing in both media. The greater the difference between a mineral's RI and that of the enclosing cement, the rougher the surface of the mineral appears to be. An arbitrary scheme used in the section of mineral descriptions is as follows:

RI	Description of relief
1.40–1.50	moderate
1.50–1.58	low
1.58–1.67	moderate
1.67–1.76	high
> 1.76	very high

The refractive indices of adjacent minerals in the thin section may be compared using the Becke line.

Alteration   The most common cause of alteration is by water or CO_2 coming into contact with a mineral, chemically reacting with some of its elements, and producing a new, stable mineral phase(s). For example, water reacts with the feldspars and produces clay minerals. In thin section this alteration appears as an area of cloudiness within the transparent feldspar grain. The alteration may

9

be so advanced that the mineral is completely replaced by a new mineral phase. For example, crystals of olivine may have completely altered to serpentine, but the area occupied by the serpentine still has the configuration of the original olivine crystal. The olivine is said to be pseudomorphed by serpentine.

1.4.2 Properties under crossed polars
The analyser is inserted into the optical path to give variably coloured grains with a dark background.

Isotropism　Minerals belonging to the cubic system are isotropic and remain dark under crossed polars whatever their optical orientation. All other minerals are anisotropic and they usually appear coloured, going into extinction (that is, going dark) four times during a complete rotation of the mineral section. However, this property varies with crystallographic orientation, and each mineral possesses at least one orientation which will make the crystal appear to be isotropic. For example, in tetragonal, trigonal and hexagonal minerals, sections cut perpendicular to the c axis are always isotropic.

Birefringence and interference colour　Under crossed polars, the colour of most anisotropic minerals varies, the same mineral showing different colours depending on its crystallographic orientation (see Plates 1b & 2b). Thus quartz may vary from grey to white, and olivine may show a whole range of colours from grey to red, blue or green. These are colours on Newton's Scale, which is divided into several orders:

Order	Colours
first	grey, white, yellow, red
second	violet, blue, green, yellow, orange, red
third	indigo, green, blue, yellow, red, violet
fourth and above	pale pinks and green

A Newton's Scale of colours can be found on the back cover of this book. These orders represent interference colours; they depend on the thickness of the thin-section mineral and the birefringence, which is the difference between the two refractive indices of the anisotropic mineral grain. The thin-section thickness is constant (normally 30 μm) and so interference colours depend on birefringence; the greater the birefringence, the higher the order of the interference colours. Since the maximum and minimum refractive indices of any mineral are oriented along precise crystallographic directions, the highest interference colours will be shown by a mineral section which has both maximum and minimum RIs in the plane of the section. This section will have the maximum birefringence (denoted δ) of the mineral. Any differently oriented section will have a smaller birefringence and show lower colours. The descriptive terms used in Chapter 2 are as follows:

Maximum birefringence (δ)	Interference colour range	Description
0.00–0.018	first order	low
0.018–0.036	second order	moderate
0.036–0.055	third order	high
> 0.055	fourth order or higher	very high

Very low may be used if the birefringence is close to zero and the mineral shows anomalous blue colours.

Interference figures   Interference figures are shown by all minerals except cubic minerals. There are two main types of interference figure (see Figs 4.9, 4.15, Plate 3), uniaxial and biaxial.

Uniaxial figures may be produced by suitably orientated sections from tetragonal, trigonal and hexagonal minerals. An isotropic section (or near-isotropic section) of a mineral is first selected under crossed polars, and then a high-power objective (× 40 or more) is used, with the substage convergent lens in position and the aperture diaphragm open. When the Bertrand lens is inserted into the optical train, a black cross will appear in the field of view. If the cross is off centre, the lens is rotated so that the centre of the cross occurs in the SW (lower left-hand) segment of the field of view.

The first-order red accessory plate is then inserted into the optical train in such a way that the length-slow direction marked on it points towards the centre of the black cross, and the colour in the NE quadrant of the cross is noted:

blue means that the mineral is positive   (denoted + ve)
yellow means that the mineral is negative   (denoted − ve)

Some accessory plates are length-fast, and the microscope may not allow more than one position of insertion. In this case the length-fast direction will point towards the centre of the black cross and the colours and signs given above will be reversed, a yellow colour meaning that the mineral is positive and a blue colour negative. It is therefore essential to appreciate whether the accessory plate is

11

length-fast or -slow, and how the fast or slow directions of the accessory plate relate to the interference figure after insertion (see Fig. 4.15).

Biaxial figures may be produced by suitable sections of orthorhombic, monoclinic and triclinic minerals. An isotropic section of the mineral under examination is selected and the microscope mode is as outlined for uniaxial figures, i.e. × 40 objective and convergent lens in position. Inserting the Bertrand lens will usually reveal a single optic axis interference figure which appears as a black arcuate line (or isogyre) crossing the field of view. Sometimes a series of coloured ovals will appear, arranged about a point on the isogyre, especially if the mineral section is very thick or if the mineral birefringence is very high. The stage is then rotated until the isogyre is in the 45° position (relative to the crosswires) and concave towards the NE segment of the field of view. In this position the isogyre curvature can indicate the size of the optic axial angle $(2V)$ of a mineral: the more curved the isogyre, the smaller the $2V$. The curvature will vary from almost a 90° angle, indicating a very low $2V$ (less than 10°) to 180° when the isogyre is straight (with a $2V$ of 80°–90°). When the $2V$ is very small (less than 10°) both isogyres can be seen in the field of view, and the interference figure resembles a uniaxial cross, which breaks up (i.e. the isogyres move apart) on rotation. The first-order red accessory plate (length-slow) is inserted and the colour noted on the *concave* side of the isogyre:

> blue means that the mineral is positive (+ ve)
> yellow means that the mineral is negative (− ve)

If the accessory plate is length-fast (as mentioned in the preceding section) the above colours will be reversed; that is, a yellow colour will be positive and blue negative (see Fig. 4.15 and Plate 3).

Extinction angle Anisotropic minerals go into extinction four times during a complete 360° rotation of a mineral section. If the analyzer is removed from the optical train while the mineral grain is in extinction, the orientation of some physical property of the mineral, such as a cleavage or trace of a crystal face edge, can be related to the microscope crosswires.

All uniaxial minerals possess *straight* or *parallel* extinction since a prism face or edge, or a prismatic cleavage, or a basal cleavage, is parallel to one of the crosswires when the mineral is in extinction (see Plate 2b).

Biaxial minerals possess either *straight* or *oblique* extinction. Orthorhombic minerals (olivine, sillimanite, andalusite and orthopyroxenes) show straight extinction against either a prismatic cleavage or a prism face edge. All other biaxial minerals possess oblique extinction, although in some minerals the angular displacement may be extremely small: for example, an elongate section of biotite showing a basal cleavage goes into extinction when these cleavages are almost parallel to one of the microscope crosswires. The angle through which the mineral then has to be rotated to bring the cleavages parallel to the crosswire will vary from nearly 0° to 9°, depending on the biotite composition; it is called the *extinction angle*.

The maximum extinction angle of many biaxial minerals is an important optical property, which has to be determined precisely. This is done as follows. A mineral grain is rotated into extinction, and the angular position of the microscope stage is noted. The polars are uncrossed (by removing the upper analyser from the optical train) and the mineral grain rotated until a cleavage trace or crystal trace edge or twin plane is parallel to the crosswires in the field of view. The position of the microscope stage is again noted, and the difference between this reading and the former one gives the extinction angle of the mineral grain. Several grains are tested, since the crystallographic orientation may vary, and the *maximum extinction angle* obtained is noted for that mineral. The results of measurements from several grains should *not* be averaged.

Extinction angles are usually given in mineral descriptions as the angle between the slow (γ) or fast (α) ray and the cleavage or face edge (written as γ or $\alpha\hat{}cl$). This technique is explained in detail in Chapter 4.

In many biaxial minerals the maximum extinction angle is obtained from a mineral grain which shows maximum birefringence such as, for example, the clinopyroxenes diopside, augite and aegirine, and the monoclinic amphiboles tremolite and the common hornblendes. However, in some minerals the maximum extinction angle is not found in a section showing maximum birefringence. This is so for the clinopyroxene pigeonite, the monoclinic amphiboles crossite, katophorite and arfvedsonite, and a few other minerals, of which kyanite is the most important (see also Ch. 4, Table 4.1).

Throughout the mineral descriptions given in Chapter 2, large variations in the maximum extinction angle are shown for particular minerals. For example, the maximum extinction angles for the amphiboles tremolite–actinolite are given as between 18° and 11° ($\gamma\hat{}$cleavage). Tremolite, the Mg-rich member, has a maximum extinction angle between 21° and 17°, whereas ferroactinolite has a maximum extinction angle ranging from 17° to 11°. This variation in the extinction angle is caused mainly by variations in the Mg:Fe ratio. Variations in extinction angles are common in many minerals or mineral pairs which show similar chemical changes.

Length-fast/-slow This property is given only where appropriate. Elongate birefringent minerals can be subdivided into the two categories, *length-fast* and *length-slow*, but this is a sensible subdivision only when the extinction angle of the mineral, relative to the direction of elongation, is relatively small. The grain is placed in extinction, with the direction of elongation oriented approximately east–west. The grain is then rotated 45° from extinction, with the elongation becoming approximately NE–SW. The interference colour of the grain is noted. A sensitive tint plate with its length-slow direction oriented NE–SW is inserted into the accessory slot and the change in the interference colour of the grain observed. If the grain is length-slow then the colour will increase by one order. If the mineral is length-fast then the colour will generally decrease by one order; minerals with low first-order grey interference colours are an exception, and it is usual to see another first-order colour produced. Rotating the grain through 90° may help to establish and confirm the slow/fast orientation of the vibration directions of the grain. The reasons for these effects are explained in Chapter 4.

Twinning This property is present when areas with differing extinction orientations within the same mineral grain have planar contacts. Often only a single twin plane is seen, but in some minerals (particularly plagioclase feldspars) multiple or lamellar twinning occurs, with parallel twin planes.

Zoning Compositional variation (zoning) within a single mineral may be expressed in terms of changes of 'natural' colour from one zone to an adjoining one (see Plate 2a), by changes in birefringence, or by changes in extinction orientation. These changes may be abrupt or gradational, and commonly occur as a sequence from the core of a mineral grain (the early-formed part) to its edge (the last-formed part).

Zoning is generally a growth phenomenon and is therefore related to the crystal shape.

Dispersion The refractive index increases as the wavelength of light decreases. Thus the refractive index of a mineral for red light is less than that for blue light (since the wavelength of red light is greater than the wavelength of blue light). White light entering a mineral section is split into the colours of the spectrum, with blue nearest to the normal (i.e. the straight-through path) and red the furthest away. This breaking up of the white light is called *dispersion*. In most minerals the amount of dispersion is very small and it will not affect the mineral's optical properties. However, the Na-rich clinopyroxenes, the Na-rich amphiboles, sphene, chlori-

14

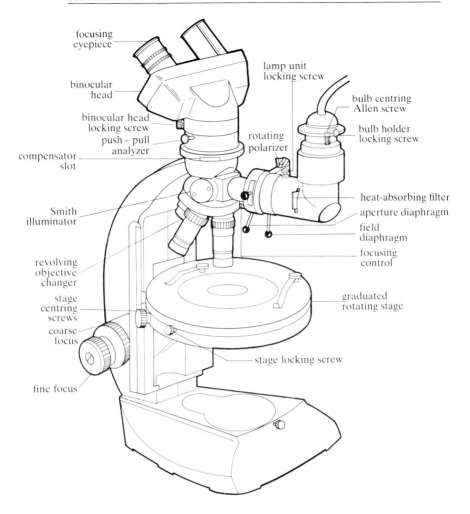

Figure 1.4 The Vickers M73 reflected-light microscope. Note that it is the *polarizer* that rotates in this microscope

toid, zircon and brookite possess very strong dispersion. With many of these minerals, interference figures may be difficult to obtain, and the use of accessory plates (to determine mineral sign, etc.) may not be possible.

Each mineral possesses a few diagnostic properties, and in the descriptions in Chapter 2 these have been marked with an asterisk. Differences between the mineral being described and other minerals that have similar optical properties are sometimes discussed in a final paragraph.

15

1.5 The reflected-light microscope

The light source A high-intensity source (Fig. 1.4) is required for reflected-light studies, mainly because of the low brightness of crossed polar images. Tungsten–halogen quartz lamps, similar to those in transparency projectors, are used and the tungsten light (A source) gives the field a yellowish tint. Many microscopists prefer to use a blue correction filter to change the light colour to that of daylight (C source). A monochromatic light source (coloured light corresponding to a very limited range of the visible spectrum) is rarely used in qualitative microscopy, but monochromatic filters for the four standard wavelengths (470 nm, 546 nm, 589 nm and 650 nm) could be useful in comparing the brightness of coexisting minerals, especially now that quantitative measurements of brightness are readily available.

The polarizer Polarized light is usually obtained by using a polarizing film, and this should be protected from the heat of the lamp by a glass heat filter. The polarizer should always be inserted in the optical train. It is best fixed in orientation to give east–west-vibrating incident light. However, it is useful to be able to rotate the polarizer on occasion in order to correct its orientation, or as an alternative to rotating the analyser.

The incident illuminator The incident illuminator sits above the objective, and its purpose is to reflect light down through the objective onto the polished specimen. As the reflected light travels back up through the objective to the eyepiece it must be possible for this light to pass through the incident illuminator. Three types of reflector are used in incident illuminators (Fig. 1.5):

(a) The cover glass or coated thin glass plate (Fig. 1.5a). This is a simple device, but it is relatively inefficient because of light loss both before and after reflection from the specimen. However, its main disadvantage when at 45° inclination is the lack of uniform extinction of an isotropic field. This is due to rotation of the vibration direction of polarized reflected light, which passes asymmetrically through the cover glass on returning towards the eyepiece. This disadvantage is overcome by decreasing the angle to about 23°, as on Swift microscopes.

(b) The mirror plus glass plate or Smith illuminator (Fig. 1.5b). This is slightly less efficient than the cover glass but, because of the low angle (approaching perpendicular) of incidence of the returning reflected light on the thin glass plate, extinction is uniform and polarization colours are quite bright. This illuminator is used on Vickers microscopes.

Figure 1.5
Incident
illuminators.

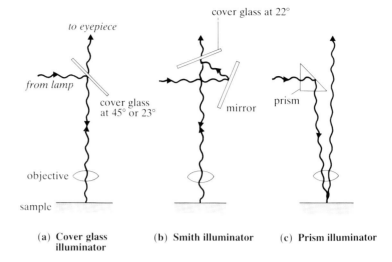

(a) **Cover glass illuminator**   (b) **Smith illuminator**   (c) **Prism illuminator**

(c) The prism or total reflector (Fig. 1.5c). This is more efficient than the glass plate type of reflector but it is expensive. It would be 100 per cent efficient, but half of the light flux is lost because only half of the aperture of the objective is used. One disadvantage is the lack of uniform extinction obtained. A special type is the triple prism or Berek prism, with which very uniform extinction is obtained (Hallimond 1970, p. 103). Prism reflectors are usually available only on research microscopes, and are normally interchangeable with glass plate reflectors. One disadvantage of prisms is that the incident light is slightly oblique, and this can cause a shadow effect on surfaces with high relief. Colouring of the shadow may also occur.

Objectives Objectives are magnifiers and are therefore described in terms of their magnification power, e.g. ×5. They are also described using numerical aperture (Fig. 1.6), the general rule being the higher the numerical aperture the larger the possible magnification. It is useful to remember that, for objectives described as being of the same magnification, a higher numerical aperture leads to finer resolved detail, a smaller depth of focus and a brighter image. Objectives are designed for use with *either* air (dry) *or* immersion oil between the objective lens and the sample. The use of immersion oil between the objective and sample leads to an increase in the numerical aperture value (Fig. 1.6). Immersion objectives are usually engraved as such.

Low-power objectives can usually be used for either transmitted or reflected light, but at high magnifications (> ×10) good images can be obtained only with the appropriate type of objective.

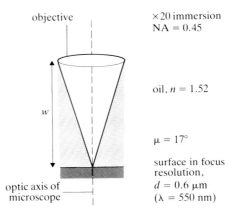

Figure 1.6 The numerical aperture and resolution. NA $= n \sin \mu$, where NA is the numerical aperture, n is the refractive index of the immersion medium, and μ is half the angle of the light cone entering the objective lens (for air, $n = 1.0$). $d = 0.5\lambda/$NA, where $d =$ the resolution (the distance between two points that can be resolved) and λ is in μm (1 μm $= 1000$ nm). The working distance (w in the diagram) depends on the construction of the lens: for the same magnification, oil immersion lenses usually have a shorter distance than dry objectives.

Reflected-light objectives are also known as metallurgical objectives. Achromatic objectives are corrected for chromatic aberration, which causes colour fringes in the image due to dispersion effects. Planochromats are also corrected for spherical aberration, which causes a loss in focus away from the centre of a lens; apochromats are similarly corrected but suffer from chromatic difference of magnification, which must be removed by the use of compensating eyepieces.

Analyser The analyser may be moved in and out of the optical train and rotated through small angles during observation of the specimen. The reason for rotation of the analyser is to enhance the effects of anisotropy. It is taken out to give plane polarized light (PPL), the field appearing bright, and put in to give crossed polars (XPOLS), the field appearing dark. Like the polarizer, it is usually made of polarizing film. On some microscopes the analyser is fixed in orientation and the polarizer is designed to rotate. The effect is the same in both cases, but it is easier to explain the behaviour of light if a rotating analyser is assumed (Section 5.3).

The Bertrand lens The Bertrand lens is little used in reflected-light microscopy, especially by beginners. The polarization figures obtained are similar to the interference figures of transmitted-light microscopy, but differ in origin and use.

Isotropic minerals give a black cross, which is unaffected by rotation of the stage but splits into two isogyres on rotation of the analyser. Lower-symmetry minerals give a black cross in the extinction position, but the cross separates into isogyres on rotation of either the stage or the analyser. Colour fringes on the isogyres relate to dispersion of the rotation properties.

Light control Reflected-light microscopes are usually designed to give Kohler-type critical illumination (Galopin & Henry 1972, p. 58). As far as the user is concerned, this means that the aperture diaphragm and the lamp filament can be seen using conoscopic light (Bertrand lens in) and the field diaphragm can be seen using orthoscopic light (Bertrand lens out).

A lamp rheostat is usually available on a reflected-light microscope to enable the light intensity to be varied. A very intense light source is necessary for satisfactory observation using crossed polars. However, for PPL observations the rheostat is best left at the manufacturer's recommended value, which should result in a colour temperature of the A source. The problem with using a decreased lamp intensity to decrease image brightness is that this changes the overall colour of the image. Ideally, neutral density filters should be used to decrease brightness if the observer finds it uncomfortable. In this respect, binocular microscopes prove less wearisome on the eyes than monocular microscopes.

Opening of the *aperture diaphragm* decreases resolution, decreases the depth of focus and increases brightness. It should ideally be kept only partially open for PPL observation, but opened fully when using crossed polars. If the aperture diaphragm can be adjusted, it is viewed using the Bertrand lens or by removing the ocular (eyepiece). The aperture diaphragm is shown correctly centred for glass plate and prism reflectors in Figure 1.7.

The *illuminator field diaphragm* is used simply to control scattered light. It can usually be focused and it should be in focus at the same position as the specimen image. The field diaphragm should be opened until it just disappears from the field of view.

1.6 The appearance of polished sections under the reflected-light microscope

On first seeing a polished section of a rock or ore sample, the observer often finds that interpretation of the image is rather difficult. One reason for this is that most students use transmitted light

Figure 1.7
Centring of the
aperture
diaphragm

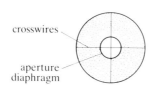

crosswires

aperture
diaphragm

**Correctly centred aperture diaphragm
for a plate glass reflector**
image with Bertrand lens inserted
and aperture diaphragm closed

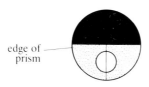

edge of
prism

**Correctly centred aperture diaphragm
for a prism reflector**
image with Bertrand lens inserted and
aperture diaphragm closed

for several years before being introduced to reflected light, and they are conditioned into interpreting bright areas as being transparent and dark areas as being opaque (see Plates 4a & b); for polished sections the opposite is the case! It is best to begin examination of a polished section such as that illustrated in Figure 1.8 by using *low-power magnification* and *plane polarized light*, under which conditions most of the following features can be observed:

(a) Transparent phases appear dark grey, because they reflect only a small proportion of the incident light, typically 3–15%. Bright patches are occasionally seen within areas of transparent minerals, and are due to reflection from surfaces under the polished surface.

(b) Absorbing phases (opaques or ore minerals) appear grey to bright white, as they reflect much more of the incident light, typically 15–95%. Some absorbing minerals appear coloured, but colour tints are usually very slight.

(c) Holes, pits, cracks and specks of dust appear black. Reflection from crystal faces in holes may give peculiar effects, such as very bright patches of light.

(d) Scratches on the polished surfaces of minerals appear as long straight or curving lines, often terminating at grain boundaries or pits. Severe fine scratching can cause a change in the appearance of minerals. Scratches on native metals, for example, tend to scatter light and cause colour effects.

(e) Patches of moisture or oil tend to cause circular dark or iridescent patches, and indicate a need to clean the polished surface.

(f) Tarnishing of minerals is indicated by an increase in colour intensity, which tends to be rather variable. Sulphides, such as

20

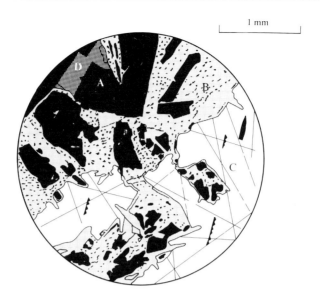

Figure 1.8 A diagrammatic representation of a polished section of a sample of lead ore. *Transparent* phases, e.g. fluorite (A), barite (B) and the mounting resin (D) appear dark grey. Their brightness depends on their refractive index. The fluorite is almost black. *Absorbing* (opaque) phases, e.g. galena (C), appear white. *Holes, pits* and *cracks* appear black. Note the black triangular cleavage pits in the galena and the abundant pits in the barite which result not from poor polishing but from the abundant fluid inclusions. *Scratches* appear as long straight or curving lines; they are quite abundant in the galena, which is soft and scratches easily.

bornite, tend to tarnish rapidly. Removal of tarnishing usually requires a few minutes of buffing and repolishing.

(g) Polishing relief, due to the differing hardnesses of adjacent minerals, causes dark or light lines along grain contacts. Small soft bright grains may appear to glow, and holes may have indistinct dark margins because of polishing relief.

1.7 Systematic description of minerals in polished section using reflected light

Most of the ore minerals described in Chapter 3 have a heading 'polished section'. The properties presented under this heading are in a particular sequence, and the terms used are explained briefly below. Not all properties are shown by each mineral, so only properties which might be observed are given in Chapter 3.

1.7.1 Properties observed using plane polarized light (PPL)

The analyser is taken out of the optical path to give a bright image (see Frontispiece).

Colour Most minerals are only slightly coloured when observed using PPL, and the colour sensation depends on factors such as the type of microscope, the light source and the sensitivity of an individual's eyes. Colour is therefore usually described simply as being a variety of grey or white, e.g. bluish-grey rutile or pinkish-white cobaltite.

Pleochroism If the colour of a mineral varies from grain to grain and individual grains change in colour on rotation of the stage, then the mineral is pleochroic. The colours for different crystallographic orientations are given when available. Covellite, for example, shows two extreme colours, blue and bluish light grey. Pleochroism can often be observed only by careful examination of groups of grains in different crystallographic orientations. Alternatively, the pelochroic mineral may be examined adjacent to a non-pleochroic mineral, e.g. ilmenite against magnetite.

Reflectance This is the percentage of light reflected from the polished surface of the mineral and, where possible, values are given for each crystallographic orientation. The eye is not good at estimating absolute reflectance but is a good comparator. The reflectance values of the minerals should therefore be used for the purpose of comparing minerals. Reflectance can be related to a grey scale of brightness in the following way (however, although followed in this book it is not a rigid scale). A mineral of reflectance ~ 15% (e.g. sphalerite) may appear to be light grey or white compared with a low-reflectance mineral (such as quartz) or dark grey compared with a bright mineral (such as pyrite):

R (%)	Grey scale
0–10	dark grey
10–20	grey
20–40	light grey
40–60	white
60–100	bright white

Bireflectance Bireflectance is a quantitative value, and for an anisotropic grain it is a measure of the difference between the maximum and minimum values of reflectance. However, bireflectance is usually assessed qualitatively, e.g.

(a)

(b)

Figure 1.9(a)   A reflected-light photomicrograph (PPL) of an elongate stibnite crystal (light grey) oriented east–west and containing an inclusion of stibnite (grey) in a different crystallographic orientation.   **(b)** As (a), but the elongate stibnite crystal has been rotated to north–south. The inclusion is now white. Stibnite exhibits a distinct bireflectance which depends on the crystallographic orientation of the section.

weak: observed with difficulty, $\Delta R < 5\%$ (e.g. hematite)
distinct: easily observed, $\Delta R > 5\%$ (e.g. stibnite, Figs 1.9a & b)

Pleochroism and bireflectance are closely related properties; the term pleochroism is used to describe change in tint or colour intensity, whereas bireflectance is used for a change in brightness.

1.7.2 Properties observed using crossed polars
The analyser is inserted into the optical path to give a dark image.

Anistropy Anistropy varies markedly with the crystallographic orientation of a section of a non-cubic mineral. It is assessed as follows:

(a) Isotropic mineral: all grains remain dark on rotation of the stage, e.g. magnetite.
(b) Weakly anisotropic mineral: slight change on rotation, seen only on careful examination using slightly uncrossed polars, e.g. ilmenite.
(c) Strongly anisotropic mineral: pronounced change in brightness and possible colour seen on rotating the stage when using exactly crossed polars, e.g. hematite.

Remember that some cubic minerals (e.g. pyrite) can appear to be anisotropic, and weakly anisotropic minerals (e.g. chalcopyrite) may appear to be isotropic. Anisotropy and bireflectance are related properties; an anisotropic grain is necessarily bireflecting, but the bireflectance in PPL is always much more difficult to detect than the anisotropy in crossed polars (see Plates 4c & d).

Internal reflections Light may pass through the polished surface of a mineral and be reflected back from below. Internal reflections are therefore shown by all transparent minerals. When one is looking for internal reflections, particular care should be paid to minerals of low to moderate reflectance (semi-opaque minerals), for which internal reflections might be detected only with difficulty and near grain boundaries or fractures. Cinnabar, unlike hematite which is otherwise similar, shows spectacular red internal reflections. (Plates 4e & f).

1.7.3 The external nature of grains
The grain shapes of minerals are determined by complex variables acting during deposition and crystallization, and subsequent recrystallization, replacement or alteration. Idiomorphic (a term used by reflected-light microscopists for well shaped or euhedral) grains are unusual, but some minerals in a polished section will be found to

have a greater tendency towards a regular grain shape than others. In the ore mineral descriptions in Chapter 3, the information given under the heading "crystals" is intended to be an aid to recognizing minerals on the basis of grain shape. Textural relationships are sometimes also given.

1.7.4   Internal properties of grains

Twinning   Twinning is best observed using crossed polars, and is recognized when areas with differing extinction orientations have planar contacts within a single grain (Plate 4d). Cassiterite is commonly twinned.

Cleavage   Cleavage is more difficult to observe in reflected light than in transmitted light, and is usually indicated by discontinuous alignments of regularly shaped or rounded pits. Galena is characterized by its triangular cleavage pits (Plate 4b). Scratches sometimes resemble cleavage traces. Further information on twinning and cleavage is given under the "crystals" heading in the descriptions of Chapter 3.

Zoning   Compositional zoning of chemically complex minerals, such as tetrahedrite, is probably very common but rarely gives observable effects such as colour banding. Zoning of micro-inclusions is more common.

Inclusions   The identity and nature of inclusions commonly observed in the mineral are given, as this knowledge can be an aid to identification. Pyrrhotite, for example, often contains lamellar inclusions of pentlandite.

1.7.5   Vickers hardness number (VHN)
The Vickers hardness number is a quantitative value of hardness, knowledge of which is useful when comparing the polishing properties of minerals (see Section 1.10).

1.7.6   Distinguishing features
Distinguishing features are given for the mineral compared with other minerals of similar appearance. The terms harder or softer refer to comparative polishing hardness (see Section 1.9).

1.8 Observations using oil immersion in reflected-light studies

Preliminary observations on polished sections are always made simply with air ($RI = 1.0$) between the polished surface and the microscope objective, and for most purposes this suffices. However, an increase in useful magnification and resolution can be achieved by using immersion objectives which require oil (use the microscope manufacturer's recommended oil, e.g. Cargille oil type A) between the objective lens and the section surface. A marked decrease in glare is also obtained with the use of immersion objectives. A further reason for using oil immersion is that the ensuing change in appearance of a mineral may aid its identification. Ramdohr (1969) states: 'It has to be emphasised over and over again that whoever shuns the use of oil immersion misses an important diagnostic tool and will never see hundreds of details described in this book.' Oil immersion nearly always results in a decrease in reflectance (Table 1.1), the reason being evident from examination of the Fresnel equation (Section 5.1.1), which relates the reflectance of a mineral to its optical properties *and* to the refractive index (N) of the immersion medium. Because it is the $n - N$ and $n + N$ values in the equation that are affected, the decrease in reflectance resulting from the increase in N is greater for minerals with a lower absorption coefficient (see Table 1.1).

Table 1.1 The relationship between the reflectances of minerals in air (R_{air}) and oil immersion (R_{oil}) and their optical constants, refractive index (n) and absorption coefficient (k). Hematite is the only non-cubic mineral represented, and two sets of values corresponding to the ordinary (o) and extraordinary (e) rays are given. N is the refractive index of the immersion medium.

		n	k	R_{air} (%) ($N = 1.0$)	R_{oil} (%) ($N = 1.52$)
Transparent minerals					
fluorite CaF_2		1.434	0.0	3.2	0.08
sphalerite ZnS		2.38	0.0	16.7	4.9
Weakly absorbing minerals					
hematite Fe_2O_3	(o)	3.15	0.42	27.6	12.9
	(e)	2.87	0.32	23.9	9.9
Absorbing (opaque) minerals					
galena PbS		4.3	1.7	44.5	28.9
silver Ag		0.18	3.65	95.1	93.2

The colour of a mineral may remain similar or may change markedly from air to oil immersion. The classic example of this is covellite, which changes from blue in air to red in oil, whereas the very similar blaubleibender covellite remains blue in both air and oil. Other properties, such as bireflectance and anisotropy, may be enhanced or diminished by the use of oil immersion.

To use oil immersion, lower the microscope stage so that the immersion objective is well above the area of interest on the horizontal polished section. Place a droplet of the recommended oil on the section surface, and preferably also on the objective lens. Slowly raise the stage using the coarse focus control, viewing from the side, until the two droplets of oil just coalesce. Continue to raise the stage very slowly using the fine focus, looking down the eyepiece until the image comes into focus. Small bubbles may drift across the field, but they should not cause any inconvenience. Larger bubbles, which tend to be caused by moving the sample too quickly, may be removed satisfactorily only by complete cleaning.

To clean the objective, lower the stage and immediately wipe the end of the objective with a soft tissue. Alcohol on a tissue may be used, but not a solvent such as acetone, which may result in loosening of the objective lens. The polished section can be carefully lifted from the stage and cleaned in the same way.

Most aspects of qualitative ore microscopy can be undertaken without recourse to oil immersion, and oil immersion examination of sections which are subsequently to be carbon coated for electron-beam micro-analysis should be avoided. The technique is most profitably employed in the study of small grains of low-reflectance materials such as graphite or organic compounds, where the benefits are a marked increase in resolution and image quality at high magnification.

1.9 Polishing hardness

During the polishing process, polished sections inevitably develop some relief (or topography) owing to the differing hardness of the component minerals (see Fig. 1.10). Soft minerals tend to be removed more easily than hard minerals. Also, the surfaces of hard grains tend to become convex, whereas the surfaces of soft grains tend to become concave. One of the challenges of the polishing technique is totally to avoid relief during polishing, because of the detrimental effect of polishing relief on the appearance of the polished section, as well as the necessity for optically flat polished surfaces for reflectance measurements. As some polishing relief is advantageous in *qualitative* mineral identification, it is often

27

Figure 1.10 A reflected-light photomicrograph (PPL). Polishing has enhanced the relief of euhedral quartz (dark grey) which is associated with coarse calcite (grey to dark grey), sphalerite (light grey) and galena (white).

beneficial to enhance the polishing relief by buffing the specimen for a few minutes using a mild abrasive such as gamma alumina on a soft nap.

Polishing relief results in a phenomenon known as the Kalb light line, which is similar in appearance to a Becke line. A sharp grain contact between a hard mineral such as pyrite and a soft mineral such as chalcopyrite should appear as a thin dark line when the specimen is exactly in focus. On defocusing slightly by increasing the distance between the specimen and objective, a fine line of bright light should appear along the grain contact in the softer mineral. The origin of this light line should easily be understood on examination of Figure 1.11. Ideally the light line should move away from the grain boundary as the specimen is further defocused. On defocusing in the opposite sense the light line appears in the harder mineral, and defocusing in this sense is often necessary, as the white line is difficult to see in a bright white soft mineral. The light line is best seen using low-power magnification and an almost closed-aperture diaphragm.

The Kalb light line is used to determine the relative polishing hardness of minerals in contact in the same polished section. This sequence can be used to confirm optical identification of the mineral set, or as an aid to the identification of individual minerals, by comparision with published lists of relative polishing hardness (e.g. Uytenbogaardt & Burke 1971).

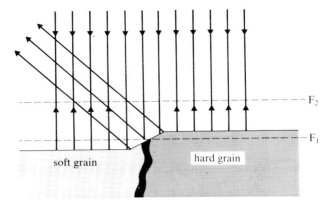

Figure 1.11   Relative polishing hardness. the position of focus is first at F_1. If the specimen is now *lowered* away from the objective, the level that is in focus will move to F_2, so that a light line (the "Kalb light line") appears to move into the *softer* substance.

Relative polishing hardness can be of value in the study of micro-inclusions in an identified host phase; comparison of the hardness of an inclusion and its surround may be used to estimate the hardness of the inclusion, or to eliminate some possibilities resulting from identification from optical properties. Similarly, if optical properties cannot be used to identify a mineral with certainty, comparison of polishing hardness with an identified co-existing mineral may help. For example, pyrrhotite is easily identified and may be associated with pyrite or pentlandite, which are similar in appearance; however, pyrite is harder than pyrrhotite, whereas pentlandite is softer.

1.10   Microhardness (VHN)

The determination of relative polishing hardness (Section 1.9) is used in the mineral identification chart (Appendix D). However, hardness can be measured quantitatively using micro-indentation techniques. The frequently used hardness value, the Vickers hardness number (VHN), is given for each mineral listed in Appendix C.

Micro-indentation hardness is the most accurate method of hardness determination and, in the case of the Vickers technique, involves pressing a small square-based pyramid of diamond into the polished surface. The diamond may be mounted in the centre of a special objective, with bellows enabling the load to be applied pneumatically (Fig. 1.12). The Commission on Ore Microscopy (COM) recommends that a load of 100 g should be applied for 15 s.

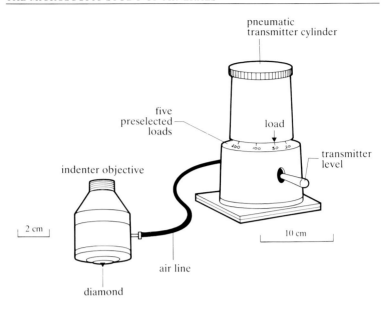

Figure 1.12 The Vickers micro-indentation hardness tester.

The size of the resulting square-shaped impression depends on the hardness of the mineral:

$$\text{VHN} = \frac{1854 \times \text{load}}{d^2} \quad \text{kg/mm}^2$$

where the load is in kilograms and d is the average length of the diagonals of the impression in microns.

Hardness is expressed in units of pressure; that is, force per unit area. Thus the micro-indentation hardness of pyrite is written:

$$\text{pyrite, VHN}_{100} = 1027\text{--}1240 \text{ kg/mm}^2$$

The subscript 100 may be omitted, as this is the standard load. As VHN values are always given in kg/mm^2, the unit is also often omitted.

The determination of hardness is a relatively imprecise technique, so an average of several indentations should be used. Tables of VHN usually give a range of values for a mineral, due to compositional variations, anisotropy of hardness, and uncertainty. Brittleness, plasticity and elasticity control the shape of the indentations and, as the shape can be useful in identification, the COM recommends that indentation shape (using the abbreviations given in Fig. 1.13) be given with VHN values.

There is a reasonable correlation between VHN and Mohs' scratch hardness, as shown in Table 1.2

30

Table 1.2 Relation between VHN and Mohs' hardness.

Mohs' hardness (H) ~ VN		
1	talc	10
2	gypsum	40
3	calcite	100
4	fluorite	200
5	apatite	500
6	orthoclase	750
7	quartz	1300
8	topaz	1700
9	corundum	2400
[10	diamond]	

1.11 Practical points on the use of the microscope (transmitted and reflected light)

Always focus using low power first. It is safer to start with the specimen surface close to the objective and *lower* the stage or raise the tube to achieve the position of focus.

Thin sections must always be placed on the stage with the cover slip on top of the section; otherwise, high-power objectives may not focus properly.

Polished samples must be level. Blocks may be mounted on a small sphere of plasticine on a glass plate and pressed gently with a levelling device. Carefully machined polished blocks with parallel faces can usually be placed directly on the stage. A level sample should appear uniformly illuminated. A more exact test is to focus on the samples, and then close the aperture diaphragm (seen using

Figure 1.13 Indentation shapes.

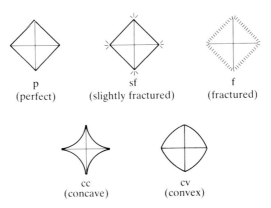

p
(perfect)

sf
(slightly fractured)

f
(fractured)

cc
(concave)

cv
(convex)

the Bertrand lens) and rotate the stage. If the sample is level, the small spot of light seen as the image should not wobble.

Good polished surfaces require careful preparation and are easily ruined. Never touch the polished surface or wipe it with anything other than a clean soft tissue, preferably moistened with alcohol or a proprietary cleaning fluid. Even a dry tissue can scratch some soft minerals. Specimens not in use should be kept covered.

The analyser is usually fixed in orientation on transmitted-light microscopes, but the polarizer may be free to rotate. There is no need to rotate the polarizer during normal use of the microscope, and it should be positioned to give east–west-vibrating polarized light. To check that the polars are exactly crossed, examine an isotropic substance such as glass and adjust the polarizer to give maximum darkness (complete extinction).

The approximate alignment of polarizer and analyzer for reflected light can be set fairly easily. Begin by obtaining a level section of a bright isotropic mineral such as pyrite. Rotate the analyzer and polarizer to their zero positions, which should be marked on the microscope. Check that the polars are crossed, i.e. the grain is dark. Rotate the analyser slightly to give as dark a field as possible. View the polarization figure (see Section 1.5). Adjust the analyzer (and/or polarizer) until a perfectly centred black cross is obtained. Examine an optically homogeneous area of a uniaxial mineral such as ilmenite, niccolite or hematite. Using crossed polars it should have four extinction positions at 90°, and the polarization colours seen in each quadrant should be identical. Adjust the polarizer and analyzer until the best results are obtained (see Hallimond 1970, p. 101).

Ensure that the stage is well centred using the high-power objective before studying optical figures.

1.12 Preparation of thin and polished sections

Thin sections are prepared by cementing thin slices of rock to glass, and carefully grinding using carborundum grit to produce a paper-thin layer of rock. The standard thickness of 30 μm is estimated using the interference colours of known minerals in the section. A cover slip is finally cemented on top of the layer of rock (Fig. 1.14).

The three common types of polished section are shown in Figure 1.14. Preparation of a polished surface of a rock or ore sample is a rather involved process which involves five stages:

(1) *Cutting* the sample with a diamond saw.
(2) *Mounting* the sample on glass or in a cold-setting resin.
(3) *Grinding* the surface, flat, using carborundum grit and water on a glass or a metal surface.

Figure 1.14
Sections.

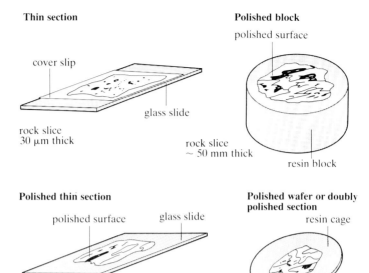

Thin section

cover slip

glass slide

rock slice
30 μm thick

Polished block

polished surface

rock slice
~ 50 mm thick

resin block

Polished thin section

polished surface   glass slide

rock slice
~ 30 μm thick

**Polished wafer or doubly
polished section**

resin cage

rock slice
50–500 μm thick

polished surfaces

3 cm

(4) *Polishing* the surface, using diamond grit and an oily lubricant on a relatively hard "paper" lap.

(5) *Buffing* the surface, using gamma alumina powder and water as lubricant on a relatively soft "cloth" lap.

There are many variants of this procedure, and the details usually depend on the nature of the samples and the polishing materials, and the equipment that happen to be available. Whatever the method used, the objective is a flat, relief-free, scratch-free polished surface. The technique used by the British Geological Survey is outlined by Lister (1978).

While *covered thin sections* continue to be popular for the study of rocks and *polished blocks* for ores, the *polished thin section* is undoubtedly the most versatile preparation, and is particularly suited to the study of samples containing a variety of minerals of low to high RI and of variable absorption (see Plates 4a & b). Variants include doubly polished thin sections, which reveal the zoning of sphalerite, and ultra-thin (preferably doubly polished) sections, which reveal textural details in fine-grained carbonates. Partially polished (to coarse diamond grade) uncovered thin sections are popular for petrographic work using cathodoluminescence microscopy. *Polished wafers* are difficult and time-consuming to prepare,

33

but are necessary for the study of fluid inclusions in transparent minerals (Shepherd et al. 1985). Examination of minerals using cathodoluminescence, ultraviolet fluorescence, lasers and electron-beam X-ray micro-analysis all require polished sections, and the use of these techniques therefore benefits from the preliminary reflected-light study of samples.

2 Silicate minerals

2.1 Crystal chemistry of silicate minerals

All silicate minerals contain silicate oxyanions $[SiO_4]^{4-}$. These units take the form of a tetrahedron, with four oxygen ions at the apices and a silicon ion at the centre. The classification of silicate minerals depends on the degree of polymerization of these tetrahedral units. In silicate minerals, the system of classification commonly used by mineralogists hinges upon how many oxygens in each tetrahedron are shared with other similar tetrahedra.

Nesosilicates   Some silicate minerals contain independent $[SiO_4]^{4-}$ tetrahedra. These minerals are known as nesosilicates, orthosilicates or *island silicates*. The presence of $[SiO_4]$ units in a chemical formula of a mineral often indicates that it is a nesosilicate, e.g. olivine $(Mg,Fe)_2SiO_4$ or garnet $(Fe,Mg \text{ etc.})_3Al_2Si_3O_{12}$, which can be rewritten as $(Fe,Mg \text{ etc.})_3Al_2[SiO_4]_3$. Nesosilicate minerals include the olivine group, the garnet group, the Al_2SiO_5 polymorphs (andalusite, kyanite and sillimanite), zircon, sphene, staurolite, chloritoid, topaz and humite group minerals.

Cyclosilicates   Cyclosilicates or *ring silicates* may result from tetrahedra sharing two oxygens, linked together to form a ring, the general composition of which is $[Si_xO_{3x}]^{2x-}$, where x is any positive integer. The rings are linked together by cations such as Ba^{2+}, Ti^{4+}, Mg^{2+}, Fe^{2+}, Al^{3+} and Be^{2+}, and oxycomplexes such as $[BO_3]^{3-}$ may be included in the structure. A typical ring composition is $[Si_6O_{18}]^{12-}$, and cyclosilicates include tourmaline, cordierite and beryl, although cordierite and beryl may be included with the tektosilicates in some classifications.

Sorosilicates   Sorosilicates contain $[Si_2O_7]^{6-}$ groups of two tetrahedra sharing a common oxygen. The $[Si_2O_7]^{6-}$ groups may be linked together by Ca^{2+}, Al^{3+}, Mg^{2+}, Fe^{2+} and some rare earth ions $(Ce^{2+}, La^{2+} \text{ etc.})$, and also contain $(OH)^-$ ions in the epidote group of minerals. In addition to the epidote group, sorosilicates include the melilites, vesuvianite (or idocrase) and pumpellyite.

Inosilicates When two or two and a half oxygens are shared by adjacent tetrahedra, inosilicates or *chain silicates* result. Minerals in this group are called single chain silicates because the $[SiO_4]^{4-}$ tetrahedra are linked together to form chains of composition $[SiO_3]_n^{2-}$, stacked together parallel to the *c* axis, and bonded together by cations such as Mg^{2+}, Fe^{2+}, Ca^{2+} and Na^+ (Fig. 2.1). Chain-silicate minerals always have a prismatic habit and exhibit two prismatic cleavages meeting at approximately right angles on the basal plane, these cleavages representing planes of weakness between chain units. The pyroxenes are single-chain inosilicates. Variations in the structure of the single chain from the normal pyroxene structure produce a group of similar, although structurally different, minerals (called the pyroxenoids, of which wollastonite is a member).

Double-chain silicates also exist in which double chains of composition $[Si_4O_{11}]_n^{6-}$ are stacked together, again parallel to the *c* crystallographic axis, and bonded together by cations such as

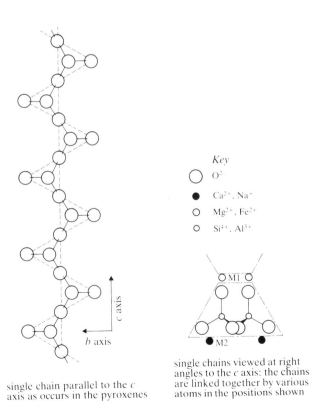

Key

○ O^{2-}

● Ca^{2+}, Na^+

○ Mg^{2+}, Fe^{2+}

○ Si^{4+}, Al^{3+}

single chain parallel to the *c* axis as occurs in the pyroxenes

single chains viewed at right angles to the *c* axis: the chains are linked together by various atoms in the positions shown

Figure 2.1
Single chain silicates.

Figure 2.2
Double chain
silicates.

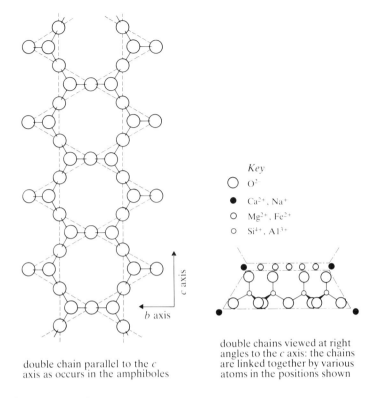

Key

○ O^{2-}

● Ca^{2+}, Na^+

○ Mg^{2+}, Fe^{2+}

○ Si^{4+}, Al^{3+}

c axis

b axis

double chain parallel to the *c* axis as occurs in the amphiboles

double chains viewed at right angles to the *c* axis: the chains are linked together by various atoms in the positions shown

Mg^{2+}, Fe^{2+}, Ca^{2+}, Na^+ and K^+, with $(OH)^-$ anions also entering the structure (Fig. 2.2). Double-chain minerals are prismatic and they possess two prismatic cleavages meeting at approximately 126° on the basal plane, these cleavages again representing planes of weakness between the double-chain units. The amphiboles are double-chain inosilicates.

Phyllosilicates When three oxygens are shared between tetra-hedra, phyllosilicates or *sheet silicates* result. The composition of such a silicate sheet is $[Si_4O_{10}]_n^{4-}$. Phyllosilicates exhibit "stacking", in which a sheet of brucite composition containing Mg^{2+}, Fe^{2+} and $(OH)^-$ ions, or a sheet of gibbsite composition containing Al^{3+} and $(OH)^-$ ions, is stacked onto an $[Si_4O_{10}]$ silicate sheet or sandwiched between two $[Si_4O_{10}]$ silicate sheets (Fig. 2.3a). Three main types exist, each of which is defined by the repeat distance of a complete multi-layered unit measured along the crystallographic axis. The 7 Å, two-layer structure includes the mineral kaolin; the 10 Å, three-layer structure includes the clay minerals montmorillonite and illite, and also the micas; and the 14 Å, four-layer structure includes chlorite. Simplified details of the main types are given in Figure 2.3b. These multilayer structures are held together by weakly

(a) Idealized tetrahedral layer of the sheet silicates

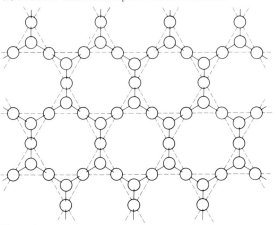

Figure 2.3
(a) Sheet silicates.
(b) The three types
of sheet silicate.

The apices of the tetrahedra all point in the same direction
(in this case upwards). Such a tetrahedral sheet may be depicted
in cross section as:

These Si–O layers are joined together by octahedral layers; either
(Al–OH) layers, called gibbsite layers and depicted by the letter G,
or (Mg, Fe–OH) layers, called brucite layers and depicted by
the letter B.

(b)

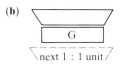

two-layer unit (1 tetrahedral layer and 1 octahedral
layer; called a 1 : 1 type)

(1) 7 Å type represented by kaolinite – serpentine is similar with a B layer replacing the G layer

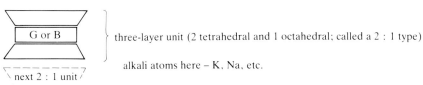

three-layer unit (2 tetrahedral and 1 octahedral; called a 2 : 1 type)

alkali atoms here – K, Na, etc.

(2) 10 Å type with muscovite, illite and montmorillonite having G octahedral layers, and biotite
B layers: the three layer units are joined together by monovalent alkali ions. Montmorillonite
may not possess any atoms in this plane and may have an overall negative charge. Water
molecules may enter the structure along these inter-unit planes

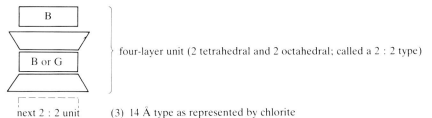

four-layer unit (2 tetrahedral and 2 octahedral; called a 2 : 2 type)

(3) 14 Å type as represented by chlorite

bonded cations (K^+, Na^+) in the micas and other 10 Å and 14 Å types. In some other sheet silicates, only Van der Waals bonding occurs between these multi-layer structures. The sheet silicates cleave easily along this weakly bonded layer, and all of them exhibit this perfect cleavage parallel to the basal plane. Minerals belonging to this group include micas, clay minerals, chlorite, serpentine, talc and prehnite.

Tektosilicates   When all four oxygens are shared with other tetra-hedra, tektosilicates or *framework silicates* form. If composed entirely of silicon and oxygen, such a framework structure will have the composition SiO_2 as in quartz. However, in many tektosilicates the silicon ion (Si^{4+}) is replaced by aluminium (Al^{3+}). Since the charges do not balance, a *coupled substitution* occurs. For example, in the alkali feldspars, one aluminium ion plus one sodium ion enter the framework structure and replace one silicon ion and, in addition, fill a vacant site. This can be written

$$Al^{3+} + Na^+ \rightleftharpoons Si^{4+} + \square \text{ (vacant site)}$$

In plagioclase feldspars a slightly different coupled substitution is required since the calcium ion is divalent:

$$2Al^{3+} + Ca^{2+} \rightleftharpoons 2Si^{4+} + \square \text{ (vacant site)}$$

This type of coupled substitution is common in the *feldspar* miner-als, and more complex substitutions occur in other tektosilicate minerals or mineral groups. Tektosilicates include feldspars, quartz, the feldspathoid group, scapolite and the zeolite group.

The classification of each mineral or mineral group is given in the descriptions in Section 2.2.

39

2.2 Mineral descriptions

This thin-section information on the silicate minerals is laid out in the same way for each mineral, as follows:

Group   Crystal chemistry
Mineral name Composition (note: Fe without superscript   Crystal system
means Fe^{2+})   axial ratio
Drawing of mineral, showing the relationship of optical to crystallographic properties
RI data
Birefringence (δ): Maximum birefringence is given for each mineral. Any variation quoted depends upon mineral composition
Uniaxial or biaxial data with sign $+$ ve (positive) or $-$ ve (negative)
Specific gravity or density   Hardness
The main properties of each mineral are then given in the following order: colour, pleochroism, habit, cleavage, relief, alteration, birefringence, interference figure, extinction angle, twinning and others (zoning etc.). Of course, only those properties which a particular mineral possesses are actually given, and **the important properties are marked with an asterisk (*)**.

Some mineral descriptions may include a short paragraph on their distinguishing features and how the mineral can be recognized from other minerals with similar optical properties.

The description ends with a short paragraph on the mineral occurrences, associated minerals and the rocks in which it is found.

Al$_2$SiO$_5$ polymorphs   Nesosilicates

Andalusite   Al$_2$SiO$_5$ orthorhombic
                                                              $0.983 : 1 : 0.704$

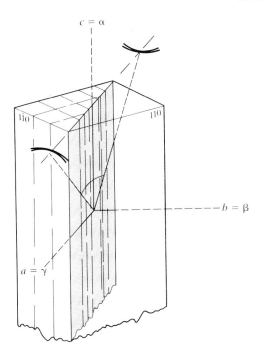

$n_\alpha = 1.629–1.649$ ⎫
$n_\beta = 1.633–1.653$ ⎪   RI variation in all polymorphs is due to ferric
$n_\gamma = 1.638–1.660$ ⎬   iron and manganese entering the structure
$\delta\ = 0.009–0.011$ ⎭
$2V_\alpha = 78°–86°$ − ve (a prism section is length fast)
OAP is parallel to (010)
$D = 3.13–3.16$ $H = 6\frac{1}{2}–7\frac{1}{2}$

COLOUR   Colourless, but may be weakly coloured in pinks.
PLEOCHROISM   Rare, but some sections show α pink, and β and γ greenish yellow.
*HABIT   Commonly occurs as euhedral elongate prisms in metamorphic
              rocks which have suffered medium-grade thermal metamorphism
              (var. **chiastolite**). Prisms have a square cross section (Fig. 2.4).
*CLEAVAGE   {110} good appearing as traces parallel to the prism edge in pris-
              matic sections but intersecting at right angles in a basal section.
RELIEF   Moderate.
ALTERATION   Andalusite can invert or change to sillimanite with increasing meta-
              morphic grade. Under hydrothermal conditions or retrograde meta-
              morphism andalusite changes to **sericite** (a type of muscovite), thus:

41

Figure 2.4 Elongate prismatic crystal of andalusite (var. chiastolite), with a basal section of the same mineral in a hornfels. The clear patches are large, spongy crystals of cordierite ($\times$ 2.5, PPL).

$$\overset{\text{from feldspar}}{} \qquad\qquad \overset{\text{sericite}}{}$$
$$3Al_2SiO_5 + 2H_2O + (3SiO_2 + K_2O) \rightarrow K_2Al_4Si_6Al_2(OH)_4O_{20}$$

*BIREFRINGENCE	Low, first order (similar to quartz).
*EXTINCTION	Straight on prism edge or on $\{110\}$ cleavages.
INTERFERENCE FIGURE	The basal section gives a Bx_a figure but $2V$ is too large to see in field of view. Look for an isotropic section and obtain an optic axis figure which will be negative.
OTHER FEATURES	Crystals in metamorphic rocks are usually poikiloblastic, and full of quartz inclusions.
OTHER TYPES	**Viridine**, or manganandalusite, is a green variety.
*OCCURRENCE	See after sillimanite.

Kyanite   Al$_2$SiO$_5$ triclinic
0.917 : 1 : 0.720
$a = 90° 5'$, $\beta = 101° 2'$, $\gamma = 105° 44'$

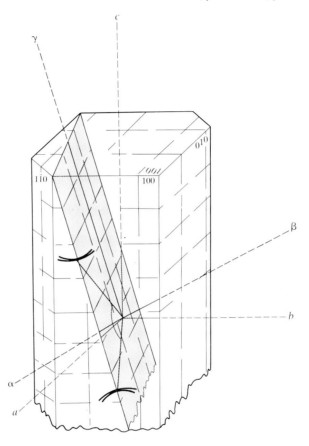

$n_\alpha = 1.712–1.718$
$n_\beta = 1.721–1.723$
$n_\gamma = 1.727–1.734$
$\delta = 0.015–0.016$
$2V_a = 82°$ − ve
OAP is approximately perpendicular to (100) with the a axis approximately the acute bisectrix
$D = 3.58–3.65$ $H = 5\frac{1}{2}–7$

COLOUR   Usually colourless in thin section, but may be pale blue.
PLEOCHROISM Weak, but seen in thick sections with a colourless, and β and γ blue.
HABIT   Usually found as subhedral prisms in metamorphic rocks. The prisms are blade-like, i.e. broad in one direction but thin in a direction at right angles to this.

43

*CLEAVAGE   {100} and {010} very good. Parting present on {001}.

*RELIEF   High: the high relief, which is easily seen if the section is held up to the light, is a very distinctive feature.

ALTERATION   As andalusite. Kyanite often occurs within large "knots" of micaceous minerals; it also inverts to sillimanite with increasing temperature.

BIREFRINGENCE   Low.

*EXTINCTION   Oblique on cleavages and prism edge; $\gamma\hat{}$ prism edge is $\sim 30°$.

INTERFERENCE FIGURE   (100) sections give Bx_a figures; but, as with andalusite, an isotropic section should be obtained and a single isogyre used to obtain the sign and size of $2V$.

TWINNING   Multiple twinning occurs on {100}.

OTHER FEATURES   The higher birefringence and excellent {100} cleavage, intersected by the {001} parting on the prism face, help to distinguish kyanite from andalusite and other index minerals.

*OCCURRENCE   See after sillimanite.

Sillimanite Al_2SiO_5 orthorhombic
0.975 : 1 : 0.752

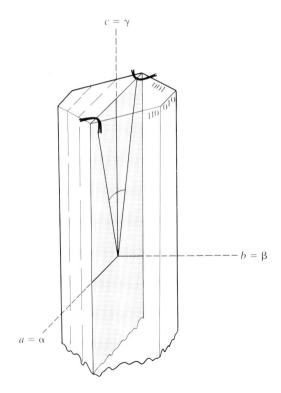

$n_a = 1.654–1.661$
$n_\beta = 1.658–1.662$

$n_\gamma = 1.678–1.683$
$\delta = 0.019–0.022$
$2V_a = 21°–30°$ + ve (a prism section is length slow)
OAP is parallel to (010)
$D = 3.23–3.27$ $H = 6\frac{1}{2}–7\frac{1}{2}$

COLOUR — Colourless.

*HABIT — It occurs as elongate prisms in two habits: either as small fibrous crystals found in regionally metamorphosed schists and gneisses, or as small prismatic crystals growing from andalusite in thermal aureoles.

*CLEAVAGE — {010} perfect: thus a basal section of sillimanite, which is diamond shaped, has cleavages parallel to the long axis.

RELIEF — Moderate.

ALTERATION — Rare.

*BIREFRINGENCE — Moderate.

*EXTINCTION — Straight on single cleavage trace.

*INTERFERENCE FIGURE — The basal section gives an excellent Bx$_a$ (+ ve) figure with a small 2V. Note that basal sections are usually small, so a very high-power objective lens will give the best figure (× 55 or more).

*OTHER FEATURES — In high-grade regionally metamorphosed rocks the fibrous sillimanite (formerly called **fibrolite**) is usually found associated with biotite, appearing as long thin fibres growing within the mica crystal (Fig. 2.5).

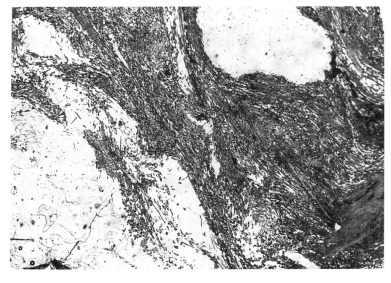

Figure 2.5 Masses of high-relief fibrous sillimanite (var. fibrolite), with clear crystals of quartz also present (× 10, PPL).

45

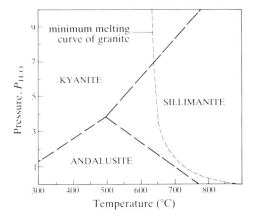

Figure 2.6 Stability relations of the three Al_2SiO_5 polymorphs. Also shown is the melting curve for albite + orthoclase + quartz + water, representing granite.

*OCCURRENCE

All three polymorphs can be used as index minerals in metamorphic rocks. They all develop in alumina-rich pelites under different conditions of temperature and pressure (Fig. 2.6). Andalusite forms at low pressures (< 1.5 kb) and low to moderate temperatures in thermal aureoles and regional metamorphism of Buchan type (high heat flow, low *P*). At higher temperatures it inverts to sillimanite. Kyanite forms at medium to high pressures and low to moderate temperatures in regional metamorphism of Barrovian type (high heat flow, moderate or high *P*). At higher temperatures kyanite also inverts to sillimanite, over a wide range of pressures and high temperatures. The sequences of mineralogical changes in pelites are:

(a) Buchan (low *P*, high heat flow ~ 60 °C/km): (low-grade) micas – andalusite (+ cordierite) – sillimanite (high-grade).

(b) Barrovian (moderate to high *P*, high heat flow ~ 30 °C/km): (low-grade) micas – staurolite – garnet – kyanite – sillimanite (highest grade).

The *P–T* diagram (Fig. 2.6) shows the stability relations of the three polymorphs. The minimum melting curve of granite has been superimposed on to the diagram. To the right (up temperature) side of this curve melting has taken place and the polymorphs would therefore occur in metamorphic rocks which had undergone some melting (e.g. migmatitic rocks).

Sillimanite can also occur in high-temperature xenoliths, found as residual products in aluminous rocks after partial melting has taken place. All of the Al_2SiO_5 polymorphs have been recognized as detrital minerals in sedimentary rocks.

Amphibole group   Inosilicates

Introduction   The amphiboles include orthorhombic and mono-clinic minerals. They possess a double-chain silicate structure which allows a large number of elemental substitutions. The double-chain has a composition of $(Si_4O_{11})_n$, with some substitution by Al^{3+} for silicon. The chains are joined together by ions that occupy various sites within the structure, and these sites are called A, X and Y. The Y sites are usually occupied by Mg^{2+} and Fe^{2+}, although Fe^{3+}, Al^{3+}, Mn^{2+} and Ti^{4+} may also enter the Y sites. The X sites are usually filled by Ca^{2+} or by Ca^{2+} and Na^+, although the ortho-rhombic amphiboles have Mg^{2+} or Fe^{2+} occupying the X sites as well as the Y ones. The A sites are always occupied by Na^+, although in the calcium-poor and calcium-rich amphiboles the A sites usually remain unoccupied.

The main amphibole groups include the following:

(a)   The Ca-poor amphiboles (Ca + Na nearly zero), which include the orthorhombic amphiboles and the Ca-poor monoclinic amphiboles. The minerals included are the anthophyllite–gedrite group (which have properties extremely similar to the cummingtonite–grunerite group in the monoclinic amphiboles).

The general formula is:

$$X_2Y_5Z_8O_{22}(OH,F)_2$$

where X = Mg,Fe, Y = Mg,Fe,Al, and Z = Si,Al.

(b)   The Ca-rich amphiboles (with Ca > Na) are monoclinic, and include the common hornblendes and tremolite–ferroactino-lite. The general formula is:

$$AX_2Y_5Z_8O_{22}(OH,F)_2$$

with A = Na (or zero in some members), X = Ca, Y = Mg,Fe,Al, and Z = Si,Al.

(c)   The alkali amphiboles are also monoclinic (with Na > Ca), and the general formula is:

$$AX_2Y_5Z_8O_{22}(OH,F)_2$$

where A = Na, X = Na (or Na,Ca), Y = Mg,Fe,Al, and Z = Si,Al. The main members are glaucophane–riebeckite, richterite and eckermannite–arfvedsonite.

The amphiboles will be examined in the above order, i.e. subgroups (a), (b) and (c), but the general optical properties of all amphibole minerals are given below:

COLOUR   Green, yellow and brown in pale or strong colours.

PLEOCHROISM   Mg-rich amphiboles may be colourless or possess pale colours with slight pleochroism, whereas iron-rich and alkali amphiboles usually are strongly coloured and pleochroic.

HABIT   Amphiboles usually occur as elongate prismatic minerals, often with diamond-shaped cross sections.

*CLEAVAGE   All amphiboles have two prismatic cleavages which intersect at 56° or 124° (Fig. 2.7).

RELIEF   Moderate to high.

ALTERATION   Common in all amphiboles; usually to chlorite or talc in the presence of water. A typical reaction is as follows:

$$Mg_2Mg_5Si_8O_{22}(OH, F)_2 + H_2O \rightarrow Mg_6Si_8O_{20}(OH)_4 + Mg(OH)_2$$
    Mg anthophyllite   talc   brucite

BIREFRINGENCE   Low to moderate: upper first-order or lower second-order interference colours occur, iron-rich varieties always giving higher interference colours. The strong colours of alkali amphiboles often mask their interference colours.

INTERFERENCE FIGURE   Apart from glaucophane and katophorite, most amphiboles have large $2V$ angles; thus an isotropic section is needed to examine a single optic axis figure. In the alkali amphiboles dispersion is so strong that interference figures may not be seen.

Figure 2.7 Several crystals of amphibole, showing the variation in brightness due to pleochroism. The grain in the centre is a basal section showing the two prismatic cleavages intersecting at 124°. The high-relief colourless crystal at the top is a euhedral section of apatite ($\times$ 10, PPL).

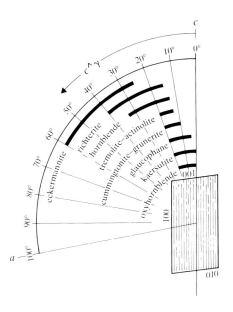

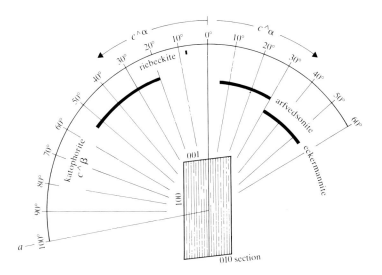

Figure 2.8 The extinction angles of amphiboles. Note that $c^\wedge\beta$ for katophorite will be the angle between the cleavage and the slow ray, since the other component in this orientation is a.

EXTINCTION   Orthorhombic amphiboles have parallel (straight) extinction. All other amphiboles are monoclinic with variable maximum extinction angles (Fig. 2.8).

ZONING   Fairly common.

TWINNING   Common on {100}; with either single or multiple twins present.

Ca-poor amphiboles

Anthophyllite $(Mg,Fe)_2(Mg,Fe)_5[Si_8O_{22}](OH,F)_2$ $Mg \gg Fe$ $\left.\rule{0pt}{18pt}\right\}$ orthorhombic

Gedrite $(Mg,Fe)_2(Mg,Fe)_3Al_2[(Si_6Al_2)O_{22}](OH,F)_2$ $Fe \gg Mg$

$1.027:1:0.292$

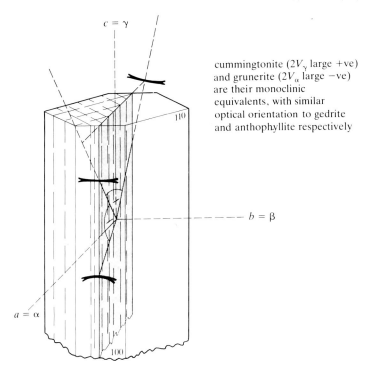

cummingtonite ($2V_\gamma$ large +ve) and grunerite ($2V_\alpha$ large −ve) are their monoclinic equivalents, with similar optical orientation to gedrite and anthophyllite respectively

$n_\alpha = 1.596–1.694$
$n_\beta = 1.605–1.710$
$n_\gamma = 1.615–1.722$
$\delta = 0.013–0.028$
$2V_\alpha = 60°–90°$ (anthophyllite) − ve $\left.\rule{0pt}{18pt}\right\}$ both crystals are length-slow
$2V_\gamma = 78°–90°$ (gedrite)   + ve
OAP parallel to (010)
$D = 2.85–3.57$ $H = 5\frac{1}{2}–6$

COLOUR   Pale brown to pale yellow.

*PLEOCHROISM	Gedrite has a stronger pleochroism than anthophyllite, with α and β pale brown, and γ darker brown.
HABIT	Elongate prismatic crystals; basal sections are recognized by inter-secting cleavages.
*CLEAVAGE	Two prismatic {110} cleavages intersecting at 54° (126°). The two cleavages are parallel to each other in a prism section and so elongate prismatic sections appear to have only one cleavage.
RELIEF	Moderate.
*ALTERATION	Common (see introduction).
BIREFRINGENCE	Low to moderate.
INTERFERENCE FIGURE	A Bx_a figure is seen on a (100) prismatic face (anthophyllite) or a basal face (gedrite), but crystals are usually so small that figures may be impossible to obtain. The best results will be obtained from a single optic axis figure.
EXTINCTION	Straight; crystals are length-slow.
OCCURRENCE	Unknown in igneous rocks, the orthorhombic amphiboles occur widely in metamorphic rocks, with anthophyllite found in associ-ation with cordierite.

Holmquistite   Holmquistite is a lithium-bearing orthoamphibole with RI = 1.62–1.67, and a moderate negative 2V. It is pleochroic, with α pale yellow, β violet and γ bluish violet, and may occur in lithium-rich pegmatites.

Cummingtonite   Cummingtonite and grunerite are the monoclinic equivalents of
Grunerite   anthophyllite and gedrite. Cummingtonite (the Mg-rich form) is positive, whereas grunerite (the Fe-rich form) is negative. 2V is large, and the density and hardness are similar to anthophyllite–gedrite. Birefringence is moderate to high (grunerite) and each mineral has oblique extinction with γˆcleavage = 10°–21° on the (010) prism face (see Fig. 2.8).

Cummingtonite occurs in metamorphosed basic igneous rocks, where it is associated with common hornblendes. Grunerite occurs in metamorphosed iron-rich sediments, where it is associated with either magnetite and quartz or with almandine garnet and fayalitic olivine, the latter minerals being common constituents of eulysite bands.

Amosite (brown asbestos) is asbestiform grunerite.

Ca-rich amphiboles

Tremolite–ferroactinolite

$$Ca_2(Mg,Fe)_5[Si_8O_{22}](OH,F)_2$$

monoclinic
$0.547:1:0.296, \beta = 105° 14'$

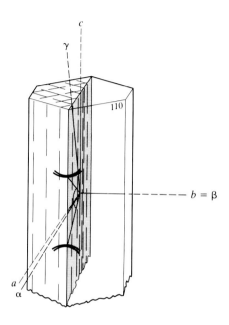

$n_\alpha = 1.599–1.688$
$n_\beta = 1.612–1.697$
$n_\gamma = 1.622–1.705$
$\delta\ = 0.027–0.017$
$2V_\alpha = 86°–65°\ -$ ve
OAP is parallel to (010)
$D = 3.02–3.44 \qquad H = 5–6$

COLOUR	Colourless to pale green (tremolite). Ferroactinolite is pleochroic in shades of green.
*PLEOCHROISM	Related to iron content – the more iron-rich, the more pleochroic the mineral – with α pale yellow, β yellowish green, and γ greenish blue.
HABIT	Elongate prismatic, with aggregates of fibrous crystals also present.
*CLEAVAGE	The usual prismatic {110} cleavages, intersecting at 56° on the basal plane.
RELIEF	Moderate to high.
ALTERATION	Common (see introduction).
BIREFRINGENCE	Moderate: second-order green is the maximum interference colour seen on a prismatic section parallel to (010).

52

INTERFERENCE FIGURE
A large $2V$ is seen on the (100) prismatic section. It is best to find an isotropic section, examine one optic axis and obtain the sign and size of $2V$.

*EXTINCTION ANGLE
The extinction angle of slow^cleavage varies from 21° to 11° depending upon the Mg:Fe ratio; the higher the ratio, the higher the extinction angle. Thus γ^cl = 21°–17° in tremolite and 17°–11° in ferroactinolite. Most amphiboles are nearly length-slow. In most thin sections, the prismatic section will rarely be correctly oriented to give a maximum extinction angle; for example, the extinction will vary from straight on a (100) section to a maximum angle on a (010) section.

*TWINNING
Amphiboles are frequently simply twinned, with {100} as the twin plane. This is shown under crossed polars by a plane across the long axis of the basal section, splitting the section into two twin halves. Multiple twinning on {100} may also occur.

OCCURRENCE
Tremolite (and actinolite) are metamorphic minerals forming during both thermal and regional metamorphism, especially from impure dolomitic limestones. At high grades tremolite is unstable, breaking down in the presence of calcite to form diopside or, in the presence of dolomite, to give olivine. Tremolite–actinolite forms during the metamorphism of ultrabasic rocks at low grades. Actinolite is a characteristic mineral of greenschist facies rocks, occurring with common hornblende, and it may also occur in blueschist rocks in association with glaucophane, epidote, albite and other minerals. Amphibolization (or uralitization) of basic igneous rocks is the name given to the alteration of pyroxene minerals to secondary amphibole by the pneumatolytic action of hydrous magmatic liquids on the igneous rocks, and the amphibole so formed may be a tremolite or actinolite.

Nephrite is the asbestiform variety of tremolite–actinolite. Precious **jade** is either nephrite or jadeite.

Hornblende series

"COMMON" HORNBLENDE
$Na_{0-1}Ca_2(Mg_{3-5}Al_{2-0})[(Si_{6-7}Al_{2-1})O_{22}](OH,F)_2$ monoclinic
$0.547:1:0.295, \beta = 105°31'$

The hornblende series is the name given to amphiboles which define a "field" of composition, the boundary end-members of which are represented by the four phases:

hastingsite $Ca_2Mg_4Al(Si_7Al)O_{22}(OH,F)_2$
tschermakite $Ca_2Mg_3Al_2(Si_6Al_2)O_{22}(OH,F)_2$
edenite $NaCa_2Mg_5(Si_7Al)O_{22}(OH,F)_2$
pargasite $NaCa_2Mg_4Al(Si_6Al_2)O_{22}(OH,F)_2$

Iron (Fe^{2+}) may replace Mg in hornblendes, but this has been omitted from the formulae for simplicity. The hornblende field can

53

(a)

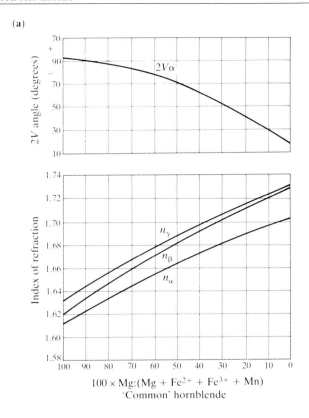

$100 \times Mg:(Mg + Fe^{2+} + Fe^{3+} + Mn)$
'Common' hornblende

(b)

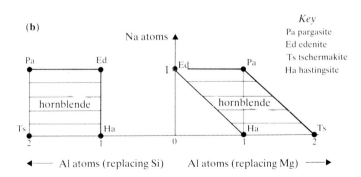

Figure 2.9 (a) The variation of the $2V$ angle and indices of refraction in the "common" hornblende series (after Deer, Howie & Zussman 1962). (b) The field of "common" hornblende compositions.

be represented on a graph by plotting the number of sodium atoms in the formulae against the number of aluminium atoms replacing either silicon or magnesium (Fig. 2.9).

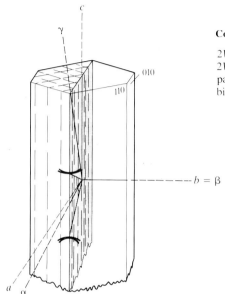

Common hornblende

$2V_\alpha$ variable, usually large, but $2V_\alpha$ for hastingsite is very small: pargasite has γ as acute bisectrix and is +ve

$\left.\begin{array}{l} n_\alpha = 1.615–1.705 \\ n_\beta = 1.618–1.714 \\ n_\gamma = 1.632–1.730 \\ \delta\ = 0.014–0.028 \end{array}\right\}$ The large variation in RI is due to compositional differences, particularly the Mg:Fe ratio in the hornblende. Ferric iron and aluminium in the Z sites will also affect both RIS and $2V$

$2V_\alpha = 15°–90°$ − ve (Mg hornblendes are + ve with $2V_\gamma$ almost 90°)
OAP is parallel to (010)
$D = 3.02–3.50$ $H = 5–6$

COLOUR	Variable, light brown or green but much darker colours for iron-rich varieties.
*PLEOCHROISM	Variable, with α pale brown or green, and β and γ brownish green. Iron-rich varieties have α yellowish brown or green, β deep green or bluish green, and γ very dark green.
HABIT	Prismatic crystals common, usually elongate.
*CLEAVAGE	Usual amphibole cleavages (see introduction). Partings parallel to {100} and {001} may also be present.
RELIEF	Moderate to high.
ALTERATION	See introductory section.
BIREFRINGENCE	Moderate: maximum interference colours are low second-order blues, but these are frequently masked by the body colour, especially if the hornblende is an iron- or sodium-rich variety.

INTERFERENCE
FIGURE
*EXTINCTION
ANGLE
TWINNING

Similar to tremolite–actinolite with extinction angle $\gamma\char`^$cl up to $30°$ (see introduction).

*OCCURRENCE   Common hornblendes are primary minerals, particularly in intermediate plutonic igneous rocks, although they can occur in other types. In intermediate rocks, the hornblende has a Fe:Mg ratio of about 1:1, whereas hornblendes are more Mg-rich in basic rocks and very iron-rich in acid rocks ($\sim 20:1$). Hornblende may occur in some basic rocks (e.g. troctolites etc.) as a corona surrounding olivine crystals, caused by reaction between olivine and plagioclase. Hornblende is stable under a wide range of pressure and temperature ($P\text{–}T$) conditions in metamorphism, being an essential constituent of the amphibolite facies. Hornblendes become more alumina rich with increasing metamorphic grade. Pure tschermakite occurs in some high-grade metamorphic rocks (often with kyanite) and pure pargasite occurs in metamorphosed dolomites. Secondary amphiboles in igneous rocks are usually tremolites or cummingtonites, but may be hornblendes.

Alkali amphiboles

Glaucophane $Na_2(Mg_3Al_2)[Si_8O_{22}](OH)_2$ monoclinic
Riebeckite $Na_2(Fe_3^{2+}Fe_2^{3+})[Si_8O_{22}](OH)_2$ $0.54:1:0.30, \beta = 104°$

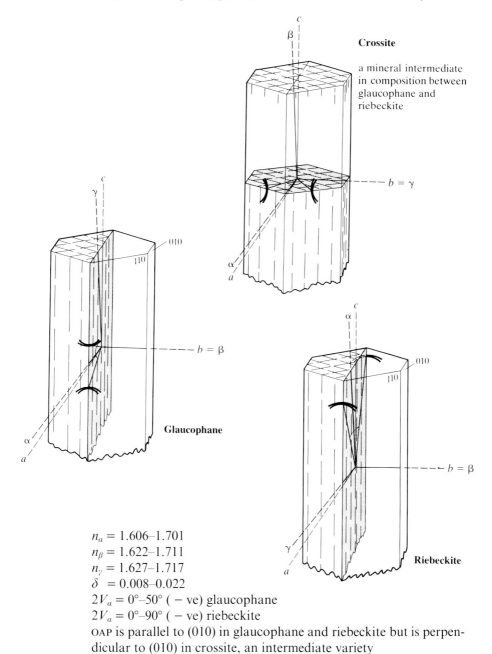

Crossite

a mineral intermediate
in composition between
glaucophane and
riebeckite

Glaucophane

Riebeckite

$n_a = 1.606–1.701$
$n_\beta = 1.622–1.711$
$n_\gamma = 1.627–1.717$
$\delta\ = 0.008–0.022$
$2V_a = 0°–50°\ (-\text{ve})$ glaucophane
$2V_a = 0°–90°\ (-\text{ve})$ riebeckite
OAP is parallel to (010) in glaucophane and riebeckite but is perpendicular to (010) in crossite, an intermediate variety
$D = 3.02–3.43$ $H = 6$ (glaucophane), 5 (riebeckite)

*COLOUR — Glaucophane is lavender blue or colourless, whereas riebeckite is dark blue to greenish.

*PLEOCHROISM — Common in both minerals, with α colourless, β lavender blue, and γ blue in glaucophane, and α blue, β deep blue, and γ yellowish green in riebeckite.

HABIT — Glaucophane usually occurs as tiny blue prismatic crystals, whereas riebeckite occurs as either large subhedral prismatic crystals or tiny crystals in the groundmass of some igneous rocks, such as alkali microgranites.

CLEAVAGE — See introduction.

RELIEF — Moderate to high.

ALTERATION — Rare in glaucophane; more common in riebeckite, which may alter to a fibrous asbestos (**crocidolite**). Riebeckite is often found in intimate association with sodic pyroxenes (aegirine), in alkali granites and syenites, for example.

BIREFRINGENCE — Low to moderate; riebeckite interference colours are usually masked by the mineral colour.

INTERFERENCE FIGURE — The optic axial angles of both minerals may vary considerably in size. In riebeckite the strong colour of the mineral may make the sign very difficult to obtain.

*EXTINCTION ANGLE — Glaucophane is length-slow with a small extinction angle of γˆcleavage (slowˆcleavage) of 6°–9°. Riebeckite is length-fast with an extinction angle of α (fast)ˆcl = 6°–8°. An (010) section in each mineral will give a maximum extinction angle. The variation in extinction angles is caused by the replacement of Al^{3+} by Fe^{3+} in glaucophane and Fe^{2+} in riebeckite.

TWINNING — Can be simple or repeated on {100}

DISTINGUISHING FEATURES — The lavender blue colour of glaucophane and the fact that it is almost length-slow, and the deep blue colour of riebeckite which is nearly length-fast, are important identification points. Where a mineral has a strong body colour, a mineral edge should be obtained which must be wedge-shaped. At the very edge the mineral is so thin that the body colour has a limited effect. Then, using a high-powered lens (e.g. × 30), whether the mineral is length-fast or length-slow can be determined using a first-order red accessory plate.

*OCCURRENCE — Glaucophane is the essential amphibole in blueschists, which form under high-P–low-T conditions in metamorphosed sediments at destructive plate margins, and are commonly found in association with ophiolite suites. Riebeckite occurs in alkali igneous rocks, especially alkali granites where it is associated with aegirine. Fibrous riebeckite (crocidolite, blue asbestos) is formed from the metamorphism at moderate T and P of massive ironstone deposits.

Richterite $Na_2Ca(Mg,Fe^{3+},Fe^{2+},Mn)_5[Si_8O_{22}](OH,F)_2$ monoclinic

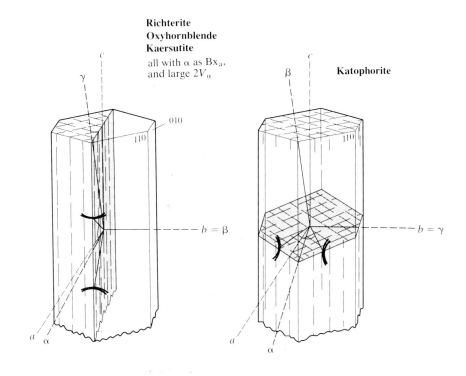

$n_\alpha = 1.605–1.685$
$n_\beta = 1.618–1.700$
$n_\gamma = 1.627–1.712$
$\delta = 0.022–0.027$
$2V_\alpha = 66°–90°$ − ve
OAP is parallel to (010)
$D = 2.97–3.45$ $H = 5\frac{1}{2}$

COLOUR   Colourless, pale yellow.
*PLEOCHROISM   Weak, in pale colours, yellow, orange and blue tints. β is usually darker in colour than α and γ, which are very pale.
HABIT   See introduction.
CLEAVAGE   Normal, see introduction.
RELIEF   Moderate to high.
BIREFRINGENCE   Moderate.
INTERFERENCE   Large $2V$ on (100) face, but an isotropic section perpendicular to a
FIGURE   single optic axis should be obtained and the sign and size of $2V$ determined from it.
EXTINCTION   Larger than normal, with a $\gamma\hat{}$cleavage of 15°–40° measured in an
ANGLE   (010) prism section.

59

TWINNING Simple or repeated on {100}.
*OCCURRENCE Rare: formed in metamorphic skarns and in thermally meta-morphosed limestones.

The following monoclinic amphiboles are also brown in colour:

Katophorite $Na_2Ca(Mg,Fe)_4Fe^{3+}[(Si_7Al)O_{22}](OH)_2$
Oxyhornblende $NaCa_2(Mg,Fe,Fe^{3+},Ti,Al)_5[(Si_6Al_2)O_{22}](O,OH)_2$
(basaltic hornblende)
Kaersutite $(Na,K)Ca_2(Mg,Fe)_4Ti[(Si_6Al_2)O_{22}(OH)_2$

COLOUR Oxyhornblende and kaersutite are dark brown in colour.
PLEOCHROISM Oxyhornblende: a yellow, and β and γ dark brown. Kaersutite: δ yellowish, β reddish brown and γ dark brownish. Katophorite is strongly coloured in yellows, browns or greens, with a yellow or pale brown, β greenish brown or dark brown, and γ greenish brown, red brown or purplish brown. In iron-rich varieties β and γ become more green and γ may be black.

INTERFERENCE All minerals are negative, with $2V_a$ of:
FIGURE

$0°–50°$ katophorite
$60°–80°$ $\begin{cases} \text{oxyhornblende} \\ \text{kaersutite} \end{cases}$

EXTINCTION Extinction angles measured on a (010) section vary with com-
ANGLE position, as follows:

$\beta\hat{}\,$cleavage $= 20°–54°$ katophorite
$\gamma\hat{}\,$cleavage $= 0°–19°$ $\begin{cases} \text{oxyhornblende} \\ \text{kaersutite} \end{cases}$

SUMMARY OF Katophorite is very strongly coloured and pleochroic in yellows,
PROPERTIES browns and greens, and with $2V_a$ variable ($0°–50°$) and a large extinction angle $\beta\hat{}\,$cl $= 20°–54°$ on a (010) section. Note that the OAP is perpendicular to (010).
　　Oxyhornblende is pleochroic in yellows and dark browns, with $2V_a$ large and with a small-angle $\gamma\hat{}\,$cl $= 0–19°$ on a (010) section.
　　Kaersutite is pleochroic in yellows and reddish browns, and with $2V_a$ large. Extinction angles are small, with $\gamma\hat{}\,$cl $= 0–19°$ on a (010) section.

OCCURRENCE Katophorite occurs in dark-coloured alkali intrusives in association with nepheline, aegirine and arfvedsonite. Kaersutite occurs in alkaline volcanic rocks, and as phenocrysts in trachytes and other K-rich extrusives, and it may be present in some monzonites.
　　Oxyhornblende occurs mainly as phenocrysts in intermediate volcanic or hypabyssal rocks such as andesites, trachytes and so on. Note that the mineral barkevikite is no longer recognized as a distinct mineral, and that the name has been formally abandoned.

Barkevikite was a name used to describe an iron-rich pargasitic hornblende, and was never chemically defined (Leake 1978).

Eckermannite–arfvedsonite

$$Na_2Na(Mg,Fe^{2+})_4Al[Si_8O_{22}](OH,F)_2 \qquad \text{monoclinic}$$

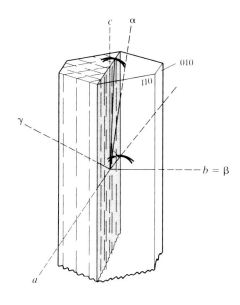 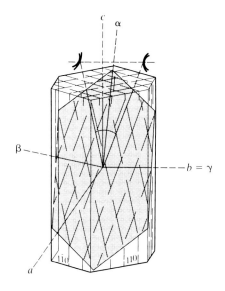

Eckermannite	Arfvedsonite
$n_a = 1.612–1.638$	$n_a = 1.674–1.700$
$n_\beta = 1.625–1.652$	$n_\beta = 1.679–1.709$
$n_\gamma = 1.630–1.654$	$n_\gamma = 1.686–1.710$
$\delta = 0.009–0.020$	$\delta = 0.005–0.012$
$2V_a = 80°–15°$ − ve	$2V_a = $ variable, probably − ve
OAP parallel to (010)	OAP perpendicular to (010)
$D = 3.00–3.16$ $H = 5\frac{1}{2}$	$D = 3.30–3.50$ $H = 5\frac{1}{2}$

COLOUR Eckermannite is pale green, and arfvedsonite has strong shades of green.

*PLEOCHROISM Eckermannite is pleochroic, with a blue green, β light green, and γ pale yellowish green; and arfvedsonite has a greenish blue to indigo, β lavender blue to brownish yellow, and γ greenish yellow to bluish grey.

HABIT Both minerals occur as large subhedral prisms, often corroded along the edges and frequently poikilitically enclosing earlier crystallizing ferromagnesian minerals.

CLEAVAGE Normal (see introduction).

RELIEF Moderate (Eck) to high (Arfv).

61

ALTERATION	Common, to chloritic minerals.
*BIREFRINGENCE	Low in both minerals, but interference colours are frequently masked by mineral colours, especially in arfvedsonite.
*INTERFERENCE FIGURE	The colour of the minerals and their strong dispersion make interference figures difficult to obtain, and these are usually indistinct, with the optic signs and the size of $2V$ impossible to judge.
*EXTINCTION	Oblique, with both minerals having variable extinction angles; $a\hat{\,}$cleavage varies from $0°$ to $50°$, but this is also difficult to obtain.
TWINNING	Simple or repeated on {100}.
OCCURRENCE	Both minerals occur as constituents in alkali plutonic rocks (soda-rich rocks), such as nepheline– and quartz–syenites, where they occur in association with aegirine or aegirine–augite and apatite. The minerals are late-crystallization products.

Aenigmatite Aenigmatite ($Na_2Fe_5^{2+}TiSi_6O_{20}$) is a mineral closely resembling the alkali amphiboles. It has very high relief (~ 1.8) and a small positive $2V$. Aenigmatite is pleochroic, with a reddish brown, β brown, and γ dark brown. It is similar to the dark brown amphiboles but has higher RIS. Aenigmatite often occurs as small phenocrysts in alkaline volcanic rocks such as phonolites.

Axinite   Cyclosilicate

Axinite $(Ca,Mn,Fe^{2+})_3Al_2(BO_3)[Si_4O_{12}](OH)$ triclinic

$0.779:1:0.978$

$a = 91°48', \beta = 98°10', \gamma = 77°18'$

$n_a = 1.674–1.693$
$n_\beta = 1.681–1.701$
$n_\gamma = 1.684–1.704$
$\delta = 0.009–0.013$
$2V_a = 63°–90° \ -ve$
OAP is perpendicular to (111)
$D = 3.26–3.36$ $H = 6\frac{1}{2}–7$

COLOUR	Colourless or pale violet brown.
PLEOCHROISM	If coloured, axinite is weakly pleochroic, with a pale yellow or brown, β violet or yellow and γ pale violet.
*HABIT	Euhedral or subhedral crystals, with wedge-shaped sections most common.
CLEAVAGE	Good on {100}; others poor.
*RELIEF	High.
ALTERATION	Uncommon to calcite or chlorite.
*BIREFRINGENCE	Low.
INTERFERENCE FIGURE	Grains will show off-centre figures with straight isogyres since $2V$ is large.

62

*EXTINCTION ANGLE	Oblique to cleavage in *all* sections.
DISTINGUISH-ING FEATURES	Axinite is difficult to identify, but wedge-shaped section, low bi-refringence, large $2V$, oblique extinction in all sections, and occurrence are important.
*OCCURRENCE	Axinite occurs in thermal aureoles between granites and carbonate sediments, along with epidote-group minerals, idocrase, grossular, diopside, datolite, tourmaline and calcite. Axinite may occur in veins or cavities in igneous rocks with prehnite, epidote, pectolite, hornblende, etc. It is a rare accessory mineral in some granite pegmatites and hydrothermal veins.

Beryl   Cyclosilicate
Beryl $Be_3Al_2[Si_6O_{18}]$ hexagonal
                                                           c/a 0.9975

$n_o = 1.560–1.602$
$n_e = 1.557–1.599$
$\delta = 0.003–0.009$
Uniaxial $-$ ve (a prism section is length fast)
$D = 2.66–2.92$ $H = 7\frac{1}{2}–8$

COLOUR	Colourless, pale yellow or pale green.
PLEOCHROISM	Weakly pleochroic in pale greens if section is thick.
*HABIT	Hexagonal prism with large basal face.
CLEAVAGE	Imperfect basal {0001}.
RELIEF	Low to moderate.
ALTERATION	Beryl easily undergoes hydrothermal alteration to clay minerals as follows, the reactions releasing quartz and **phenakite**:

$$2Be_3Al_2Si_6O_{18} + 4H_2O \rightarrow Al_4Si_4O_{10}(OH)_8 + 5SiO_2 + 2Be_2SiO_4$$
$$\qquad\qquad\qquad\qquad\qquad kaolin \qquad\qquad\qquad phenakite$$

BIREFRINGENCE	Low first order greys.
TWINNING	Rare.
*OCCURRENCE	Beryl occurs in vugs and granites and particularly in pegmatites, often associated with cassiterite. The precious stone variety, aqua-marine, occurs in similar locations, but emerald is usually found in metamorphic biotite schists.

Chlorite group   Phyllosilicates
Introduction The chlorite group of minerals are closely related to the micas, but have a 2:2 sheet structure, compared to the micas which have a 2:1 sheet structure (see p. 107). There are different types of chlorites depending upon their composition: in particular, the Mg-rich chlorites, which form a series from **amesite** to **antigorite**, including **chlinochlore** and **penninite**; and the ferrous-rich chlorites

63

which include **chamosite**. Some of the chlorites are termed *septichlorites*: these include amesite and chamosite, and are closely related to the chlorites, being different only in having a 7 Å structure rather than the normal chlorite 14 Å structure. The chlorite minerals are all similar in their optical properties, except that the iron-rich varieties (especially the ferric-rich ones) have stronger pleochroic schemes, usually in browns and yellows.

Chlorite $(Mg,Al,Fe)_{12}[(Si,Al)_8O_{20}](OH)_{16}$ monoclinic
$0.58:1:1.53, \beta = 96°17'$

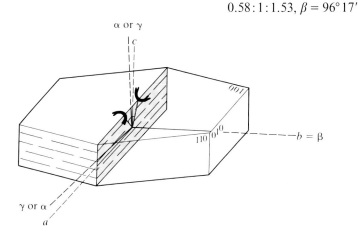

$n_\alpha = 1.57–1.66$
$n_\beta = 1.57–1.67$
$n_\gamma = 1.57–1.67$
$\delta = 0.0–0.01$
$2V = 20°–60°$ + ve or − ve
OAP is parallel to (010)
$D = 2.6–3.3$ $H = 2–3$

*COLOUR   Colourless or green (see Plate 1a).
PLEOCHROISM   Green varieties have a pale green to colourless, and β and γ darker green.
HABIT   Tabular crystals with a pseudo-hexagonal shape.
*CLEAVAGE   Perfect {001} basal cleavage.
RELIEF   Low to moderate.
ALTERATION   Oxidation of iron in chlorite may occur (the sign changes from + ve to − ve).
*BIREFRINGENCE   Very weak, usually with anomalous deep *Berlin blue* colour (see Plate 1b).
INTERFERENCE   Biaxial Bx_a figure on basal section with small $2V$. Usually positive
FIGURE   but some varieties – chamosite in particular – are optically negative. Interference figures are rarely obtained.

EXTINCTION   Straight to cleavage, but can be oblique with small-angle γ or $a\char`^cl$ (fast or slow to cleavage); very small angle ($< 5°$) on (010) section.

TWINNING   As in micas: rare.

OCCURRENCE   Chlorite is a widely distributed primary mineral in low-grade regional metamorphic rocks (greenschists), eventually changing to biotite with increasing grade; muscovite is also involved in the reaction. The initial material is usually argillaceous sediments, but basic igneous rocks and tuffs will give chlorite during regional metamorphism. In some alkali-rich rocks, chlorite will break down with increasing P and T and help to form amphibole and plagioclase. In igneous rocks chlorite is usually a secondary mineral, forming from the hydrothermal alteration of pyroxenes, amphiboles and biotites. Chlorite may be found infilling amygdales in lavas with other minerals, and may occur as a primary mineral in some low-temperature veins.

Chlorites are common in argillaceous rocks, where they frequently occur with clay minerals, particularly illite, kaolin and mixed-layer clays.

Chloritoid   Nesosilicate

Chloritoid (ottrelite)

$(Fe,Mg)_2(Al,Fe^{3+})Al_3O_2[SiO_4]_2(OH)_4$ monoclinic

$1.725:1:3.314, \beta = 101°30'$

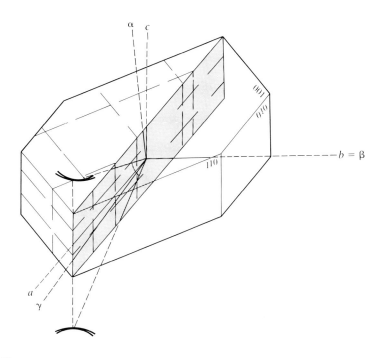

$n_\alpha = 1.713–1.730$
$n_\beta = 1.719–1.734$
$n_\gamma = 1.723–1.740$
$\delta\ = 0.010$
$2V_\gamma = 45°–68°$ + ve (normal range). $2V$ can be highly variable, with
$\quad 2V_\gamma\ 36°–90°$ + ve and $2V_\alpha\ 90°–55°$ − ve
OAP is parallel to (010)
$D = 3.51–3.80 \qquad H = 6\frac{1}{2}$

COLOUR	Colourless, green, blue green.
*PLEOCHROISM	Common, with α pale green, β blue and γ colourless to pale yellow.
HABIT	Closely resembles mica minerals, occurring as pseudo-hexagonal tabular crystals.
*CLEAVAGE	{001} perfect. Another poor, possibly prismatic, fracture may be present, and may distinguish chloritoid from micaceous minerals.
*RELIEF	High.
ALTERATION	Chloritoid may alter to muscovite and chlorite, but this is not common.
BIREFRINGENCE	Low, but masked by the greenish colour of the mineral: anomalous blue colours are often seen.
INTERFERENCE FIGURE	A bluish green coloured (100) section of chloritoid will give a Bx_a figure with a moderate $2V$ and positive sign.
EXTINCTION	Straight to perfect on {001} cleavage.
*ZONING	Zoning occasionally appears as a peculiar hourglass shape seen on prismatic sections.
OCCURRENCE	Chloritoid occurs in regionally metamorphosed pelitic rocks with a high $Fe^{3+} : Fe^{2+}$ ratio, at low grades of metamorphism. Chloritoid develops at about the same time as biotite, changing to staurolite at higher grades. Chloritoid can occur in stress-free environments, where it usually shows triclinic crystal form, particularly in quartz carbonate veins and in altered lava flows.

Clay minerals   Phyllosilicates

Introduction The clay minerals are a poorly defined group of secondary minerals, formed near or on the Earth's surface by either hydrothermal alteration or weathering processes of existing minerals, particularly feldspars and other aluminosilicates. They are usually very fine grained and difficult to identify by optical methods, so that precise identification of the clay type is normally made by X-ray diffraction techniques (XRD), scanning electron microscopy (SEM), or by analysis using a microprobe. Most clay minerals are phyllosilicates, with sheet structures containing brucite-type layers $[Mg(OH)_2]$ or gibbsite-type layers $[Al(OH)_3]$ of cations in octahedral co-ordination, and $[Si_4O_{10}]$ layers of Si^{4+} and Al^{3+} ions in tetrahedral coordination. There are two distinct groups:

(a) those with a brucite or gibbsite layer coupled to a single $[Si_4O_{10}]$ layer, called 1:1 sheet silicates (or 1:1 layered structures);

(b) those with a brucite or gibbsite layer coupled to two $[Si_4O_{10}]$ layers, called 2:1 sheet silicates (or 2:1 layered structures).

Note that chlorites and related minerals possess 2:2 layered structures where two brucite or gibbsite layers are joined to two $[Si_4O_{10}]$ layers. (See p. 38 for a fuller explanation of these structures.)

The various groups of clay minerals include:

Kaolin group, or kandites (1:1 structures)
  kaolinite
  dickite
  nacrite
  halloysite
  allophane (amorphous kaolinite)
Montmorillonite group, or smectites (2:1 structures)
  montmorillonite
  beidellite
  nontronite
Illite group, or hydromicas (2:1 structures)
  illite
  glauconite (described with the micas, p. 111)
Vermiculite group
Palygorskite group

Vermiculites are clays closely related to the montmorillonites and the chlorites, and palygorskites (sometimes called attapulgites) are fibrous clays.

As clays are rarely identified by optical means, only brief descriptions of the optical properties of the main clay groups are given below, with a statement on their main occurrences at the end of the montmorillonite section.

Kaolinite (kandite)

$Al_4[Si_4O_{10}](OH)_8$ 　　　　　　　　　triclinic

0.576 : 1 : 0.830

$a = 91°48'$. $\beta = 104°30'$, $\gamma = 90°$

$n_\alpha = 1.553–1.565$
$n_\beta = 1.56–1.57$
$n_\gamma = 1.56–1.57$
$\delta = 0.006$
$2V_\alpha = 24°–50°$ − ve
OAP perpendicular to (010)
$D = 2.61–2.68$ 　　　$H = 2–2\frac{1}{2}$

COLOUR	Colourless.
*HABIT	Similar to mica group, but crystals are extremely tiny.
RELIEF	Low.
CLEAVAGE	Perfect basal – similar to micas.
BIREFRINGENCE	Low, greys of first order.
INTERFERENCE FIGURE	The size of individual crystals is such that interference figures can rarely be obtained.
EXTINCTION	Straight, but occasional slight extinction angle on (010) face.

Illite $K_{1-1.5}Al_4[Si_{7-6.5}Al_{1-1.5}O_{20}](OH)_4$ monoclinic

$n_\alpha = 1.54–1.57$
$n_\beta = 1.57–1.61$
$n_\gamma = 1.57–1.61$
$\delta = 0.03$
$2V = $ small $(< 10°)$ − ve
OAP approx. parallel to (010)
$D = 2.6–2.9$ $H = 1–2$

The properties are similar to those of kaolin, with the exception of the birefringence, which is much stronger with second-order colours.

Montmorillonite group (smectites)

$(\frac{1}{2} Ca,Na)_{0.7}(Al,Mg,Fe)_4[(Si,Al)_8O_{20}](OH)_4.n\,H_2O$ monoclinic

$n_\alpha = 1.48–1.61$
$n_\beta = 1.50–1.64$
$n_\gamma = 1.50–1.64$
$\delta = 0.01–0.04$
$2V_\alpha$ variable, $0°$ to quite large − ve
OAP is parallel to (010)

Properties are similar to those of kaolin and illite.

OCCURRENCE OF CLAYS	Kaolin is the most common of the clay minerals, and it forms by hydrothermal alteration or weathering of feldspars, feldspathoids and other silicates. Kaolin, therefore, usually forms from the alteration of acid igneous rocks (granites, etc.), with non-alkaline conditions being required.

Illite is the common clay mineral in clays and mudstones, and it is formed by weathering of feldspars or by alteration of other clay minerals during sediment formation. Illite formation is favoured by alkaline conditions and high Al and K activities.

Montmorillonite and **beidellite** (an Mg-bearing montmorillonite) are principal constituents of bentonite clays, formed from alteration of pyroclastic ash deposits (tuff etc.). Montmorillonite (particularly Fuller's earth) is formed by alteration of basic igneous rocks in areas of poor drainage when Mg is not removed. An alkaline environment is preferred, with low K and higher Ca. Vermiculite is another clay mineral derived from biotite alteration, and with properties closely related to the smectite group.

Cordierite

Cordierite
$Al_3(Mg,Fe)_2[Si_5AlO_{18}]$

Cyclosilicate
orthorhombic
pseudo-hexagonal
$1.748:1:0.954$

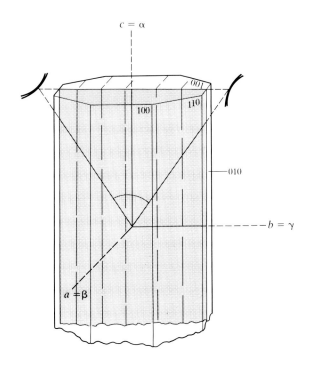

$n_\alpha = 1.522–1.558$
$n_\beta = 1.524–1.574$
$n_\gamma = 1.527–1.578$
$\delta\ = 0.005–0.020$
$2V_\gamma = 65°–90°\ +$ ve
$2V_\alpha = 90°–76°\ -$ ve
OAP is parallel to (100)
$D = 2.53–2.78$ $H = 7$

COLOUR   Colourless, but occasionally blue. The precious variety of cordierite (water sapphire) is pale violet and pleochroic.

PLEOCHROISM Iron-rich varieties show α colourless, and β and γ violet, whereas magnesium-rich varieties show pleochroism only in thick sections, with α pale yellowish green, and β and γ pale blue.

*HABIT In regionally metamorphosed rocks, cordierite occurs as large "spongy" crystals containing inclusions of muscovite, biotite and quartz. Almost all of the crystal shows alteration, and no good crystal face edges occur. In thermal aureoles, cordierite occurs in inner zones, and the fresh cordierite crystals which develop often show good hexagonal crystal form. Cordierite may occur in some igneous rocks, where it shows subhedral to euhedral crystal form.

CLEAVAGE {010} good, with a poor {001} basal cleavage sometimes developing.

*RELIEF Low, similar to quartz; usually higher than 1.54.

*ALTERATION Cordierite shows alteration at edges and along cracks to **pinite** (a mixture of fine muscovite and chlorite or serpentine), which usually appears as a pale yellowish-green mineral in thin section. A yellow pleochroic halo may sometimes appear in a cordierite crystal, surrounding an inclusion of zircon or monazite, similar to those found in biotite crystals. Such haloes are caused by the elements of the radioactive series U–Ra and Th–Ac.

BIREFRINGENCE Low, similar to quartz or feldspar.

INTERFERENCE FIGURE $2V$ is very large, and so the best figure would be obtained by examining a near isotropic section giving an optic axis figure. Such a figure is approximately found in the position of face (011) or (0$\bar{1}$1).

*TWINNING Extremely common in all crystals except those in regional metamorphic rocks where, in any case, alteration masks any twinning. Fresh, clear crystals in thermal aureoles, and some "partial melt" igneous rocks, show two kinds of twinning – cyclic and lamellar. Cyclic twinning on {110} or {130} produces a pseudo-trigonal or pseudo-hexagonal pattern (Fig. 2.10). In some instances, twinning on these planes produces lamellar twinning that is similar to twinning seen in plagioclase feldspars.

*OCCURRENCE Cordierite is a mineral found in pelitic rocks which have been subjected to metamorphism at low pressure. Cordierite occurs in the inner (high-temperature) zone of thermal aureoles, and in regional metamorphic conditions of high heat flow and low pressure, such as Buchan-type metamorphism, where the sequence of index minerals produced under regional conditions is, progressively: biotite—andalusite—cordierite—sillimanite. In these rocks, cordierite occurs in high-grade gneisses, either under abnormal P–T conditions or where a thermal metamorphic episode follows regional metamorphism, and pre-existing minerals such as kyanite and biotite become unstable, reacting to give cordierite and muscovite.

Figure 2.10 A large, spongy crystal of cordierite, showing complex sector twinning (× 2.5, XPOLS).

Cordierite may occur in some igneous rocks, especially cordierite norites. These were originally considered to represent the crystallized products of basic magma contaminated by the assimilation of argillaceous material. However, it has recently been suggested that basic igneous intrusions, emplaced into argillaceous rocks, may develop extremely high temperatures in their innermost thermal aureole such that the argillaceous material partially melts. On crystallizing, this partially melted material gives rise to cordierite norites which do not possess any magmatic component. Cordierite has also been known to occur as a primary mineral in some granites and granite pegmatites.

Datolite   Nesosilicate

Datolite   CaB[SiO₄]OH

$CaB[SiO_4]OH$

monoclinic
$1.264:1:0.632, \beta = 90°09'$

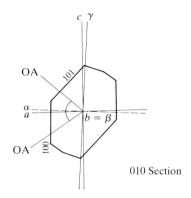

010 Section

$n_a = 1.622{-}1.626$
$n_\beta = 1.649{-}1.654$
$n_\gamma = 1.666{-}1.670$
$\delta\ = 0.044{-}0.046$
$2V_a = 72°{-}75°\ -\text{ve}$
OAP is perpendicular to b axis
$D = 2.9{-}3.0 \qquad H = 5{-}5\tfrac{1}{2}$

COLOUR	Colourless.
HABIT	Euhedral crystals are common, with complex crystal forms.
*CLEAVAGE	None.
RELIEF	Moderate.
*BIREFRINGENCE	Strong, up to middle third order colours.
INTERFERENCE FIGURE	Large $-2V$ seen on (100) sections; an optic axis is approximately perpendicular to (101).
DISTINGUISH-ING FEATURES	Datolite has no cleavage, and has higher birefringence than zeolites and other associated minerals.
OCCURRENCE	Datolite occurs in vugs and cavities of basalts along with zeolites, prehnite and calcite. It occurs in contact limestone skarns with calcite, grossular, prehnite and fluorite.

72

Dumortierite   Nesosilicate

Dumortierite $(Al,Fe^{3+})_7O_3(BO)_3[SiO_4]_3$ orthorhombic
                                                                0.583 : 1 : 0.234

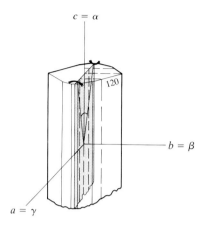

$n_\alpha = 1.655–1.686$
$n_\beta = 1.675–1.722$
$n_\gamma = 1.684–1.723$ } RIS increase with increasing Fe^{3+}
$\delta = 0.011–0.027$
$2V_a = 15°–52°$ − ve
OAP is parallel to (010) (a prism section is length fast)
$D = 3.26–3.41$ $H = 7–8\frac{1}{2}$

*COLOUR   Brightly and vividly coloured in violet, brown, green and blue.
*PLEOCHROISM   Strongly pleochroic, with α deep blue, bluish violet, green and
               brown, β colourless, greenish yellow, yellow and pale violet, and γ
               colourless, greenish yellow and pale blue.
HABIT   Fibrous or columnar aggregates of crystals are common.
CLEAVAGE   Distinct cleavage on {100}.
RELIEF   High.
ALTERATION   Dumortierite alters readily to white mica, and changes to **mullite**
               ($3Al_2O_3.2SiO_2$) at high temperatures.
BIREFRINGENCE Maximum birefringence gives high first-order colours.

INTERFERENCE   Basal sections give good Bx_a centred figures with moderate $2V$.
FIGURE
OCCURRENCE   Dumortierite occurs in late-stage acid igneous rocks with tourma-
               line, apatite, xenotime, monazite, zircon, rutile, etc. It is found in
               granite gneisses and mica schists with cordierite and the Al_2SiO_5
               polymorphs.

Epidote group   Sorosilicates

a-zoisite has *ac* as the optic axial plane.
β-zoisite has *ab* as the optic axial plane.
All monoclinic epidotes have *ac* as the optic axial plane.

Minerals in the epidote group belong to both the orthorhombic and
the monoclinic systems. Conventionally, the mineral belonging to
the higher symmetry system (orthorhombic) is described first, and
therefore the descriptions begin with zoisite and go on to the two
important monoclinic varieties, clinozoisite and epidote. Two other
important varieties include **piemontite**, an Mn-rich epidote possessing spectacular pleochroism, with *a* yellow, *β* purplish blue and *γ*
red, and found in manganiferous metamorphic rocks; and **allanite**,
or orthite, a cerium-bearing epidote with yellow to reddish-brown
pleochroism, which occurs as an accessory mineral in syenites.

Zoisite $Ca_2Al_3[Si_3O_{12}](OH)$ orthorhombic
                                                            $2.879:1:1.791$

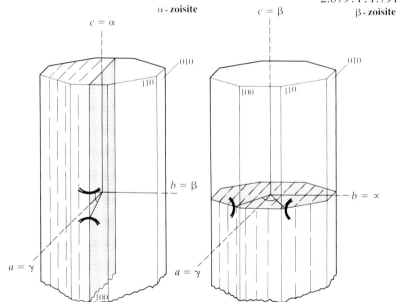

$n_\alpha = 1.685–1.707$
$n_\beta = 1.688–1.711$ } The RIS may vary depending on the amount of
$n_\gamma = 1.697–1.725$ trace elements (Fe^{3+} etc.) in the structure
$\delta = 0.005–0.020$
$2V_\gamma = 0°–60°$ + ve
OAP is either parallel to (010) in *a*-zoisite, or parallel to (001) in
β-zoisite
$D = 3.15–3.36$ $H = 6$

74

COLOUR   Colourless; a pink variety (thulite) may occur in Mn-rich environments.

HABIT   Usually found in clusters of elongate prismatic crystals, with rectangular cross sections.

CLEAVAGE   Perfect {100} prismatic cleavage, poor {001} cleavage sometimes present.

*RELIEF   High.

ALTERATION   Since zoisite forms under conditions of low P and T, it remains stable and is not subject to further reactions. Zoisite may be produced from the breakdown of calcium plagioclase from basic igneous rocks which have suffered hydrothermal alteration (a process called saussuritization).

*BIREFRINGENCE   Very low (varies between 0.004 and 0.008). a-zoisite shows low first-order colours (greys, whites), but β-zoisite shows anomalous interference colours of a deep Berlin blue.

EXTINCTION   Straight on prism edge or {100} cleavage.

*INTERFERENCE FIGURE   A (10$\bar{1}$) section gives a biaxial positive figure with a moderate $2V$.

OTHER FEATURES   Since β-zoisite differs from a-zoisite in containing up to 5% Fe_2O_3, an intermediate variety between a- and β-zoisite may occur, which is distinguished by possessing a very small $2V$ ($\approx 0°$).

OCCURRENCE   Zoisite is found in basic igneous rocks which have been hydrothermally altered, where it develops from calcic plagioclase. It also occurs in medium-grade metamorphosed schists in association with sodic plagioclase, amphibole, biotite and garnet. It may occur in some metamorphosed impure limestones.

Clinozoisite $Ca_2Al_3[Si_3O_{12}](OH)$ monclinic
                    $1.583:1:1.814, \beta = 115°30'$

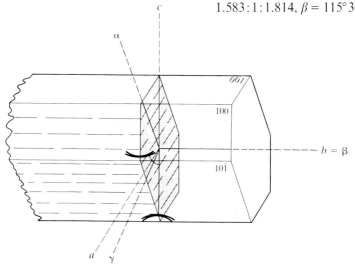

$n_a = 1.703–1.715$
$n_\beta = 1.707–1.725$ } RIS are based on a clinozoisite, with about 1% Fe_2O_3 present. These will increase with
$n_\gamma = 1.709–1.734$ increasing Fe_2O_3 content or decrease if clinozoisite is iron free

$\delta = 0.004–0.012$ (variable with Fe_2O_3)
$2V_\gamma$ = variable, usually $14°–90°$ + ve
OAP is parallel to (010)
$D = 3.12–3.38$ $H = 6\frac{1}{2}$

COLOUR	Colourless.
HABIT	Found in columnar aggregates of crystals, which are usually quite small.
CLEAVAGE	Perfect {001} cleavage, appearing as a prismatic cleavage in sections, since the mineral is elongate parallel to the b axis.
*RELIEF	High.
ALTERATION	None.
*BIREFRINGENCE	Very low with anomalous first-order interference colours (deep blue, greenish yellow; no first-order white).
EXTINCTION	Oblique, variable extinction angles depending on mineral composition, but most elongate prismatic sections have straight extinction on cleavage.
*INTERFERENCE FIGURE	A (100) section will give a biaxial positive figure, but since $2V$ is large an isotropic section should be selected and a single isogyre examined for sign and size.
OCCURRENCE	Clinozoisite occurs primarily in regionally metamorphosed low-grade rocks forming from micaceous minerals. Its other occurrences are similar to those of zoisite.

Epidote (pistacite)

$$Ca_2Fe^{3+}Al_2[Si_3O_{12}](OH)$$

monoclinic

$1.591:1:1.812, \beta = 115°24'$

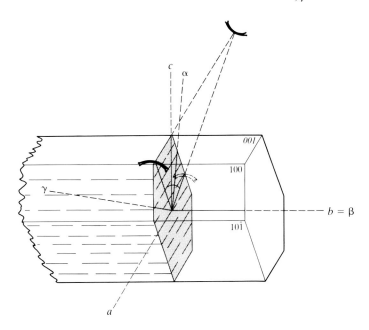

$n_\alpha = 1.715–1.751$
$n_\beta = 1.725–1.784$ } RIS increase with increasing Fe^{3+} content
$n_\gamma = 1.734–1.797$
$\delta = 0.012–0.049$ (variable with composition)
$2V_\alpha$ = variable, $64°–90° -$ ve
OAP is parallel to (010)
$D = 3.38–3.49 \qquad H = 6$

COLOUR	Colourless to pale yellowish green.
*PLEOCHROISM	Slightly pleochroic, with α colourless to pale yellowish green, β greenish and γ yellowish.
HABIT	Found in aggregates of elongate prismatic crystals with pseudo-hexagonal cross sections.
CLEAVAGE	{001} perfect (similar to clinozoisite).
*RELIEF	High.
ALTERATION	None.
*BIREFRINGENCE	Moderate to high, showing low second-order to upper third-order colours. Some sections may show low anomalous interference colours similar to clinozoisite.
EXTINCTION	Oblique to cleavage in pseudo-hexagonal sections (see figure); otherwise straight on cleavage in prismatic section.

INTERFERENCE
FIGURE
A section perpendicular to the c axis will give a biaxial negative figure with a large $2V$; thus an optic axis figure is preferable when determining the sign.

OTHER
FEATURES
Lamellar twinning may sometimes be present.

*OCCURRENCE
Epidote is an important mineral in low-grade regional metamorphic rocks, where it marks the beginning of the epidote–amphibolite facies, forming from the breakdown of chlorite. Epidote also forms from saussuritization of plagioclase feldspar and from the breakdown of amphiboles in basic igneous rocks, these changes being due to late-stage hydrothermal alteration. In highly amphibolitized basic igneous rocks, clusters of epidote crystals are commonly seen in association with plagioclase feldspar, and often within amphibole minerals.

Feldspar group   Tektosilicates

Introduction The feldspars are the most important minerals in rocks. They occur in igneous, metamorphic and sedimentary rocks, and their range of composition has led to their use as a means of classifying igneous rocks. Feldspars are absent only from certain ultramafic and ultra-alkaline igneous rock types and carbonatites. Feldspars occur in almost all metamorphic rocks, being absent only from some low-grade pelitic types, pure marbles and pure quartzites. In sedimentary rocks feldspars are common constituents of many arenaceous rocks, but are less common in shales (clay rocks) and mudstones. Feldspars are difficult to detect (although XRD investigation would reveal their presence) because of the minute grain size (< 0.002 mm) of these argillaceous rocks.

Although the most ubiquitous of minerals, feldspars have a restricted range of composition. There are two main types of feldspar:

(a) alkali feldspars, which range beween the end members orthoclase $KAlSi_3O_8$ and albite $NaAlSi_3O_8$.
(b) plagioclase feldspars, which range between the end members albite $NaAlSi_3O_8$ and anorthite $CaAl_2Si_2O_8$.

From this it is obvious that albite is common to both feldspar types, and the most usual way of depicting the complete feldspar group is in a ternary (or triangular) diagram, with orthoclase, albite and northite representing the composition of each apex (Fig. 2.11). This reveals that the alkali feldspars can contain up to 10% anorthite molecule in their structure, and similarly plagioclase feldspars can contain up to 10% orthoclase molecule in their structure. In Figure 2.11, the plagioclase feldspars are divided into six compositional divisions, with **albite**, having a composition of 100% to 90% of the

78

Figure 2.11
The feldspar
composition
diagram.

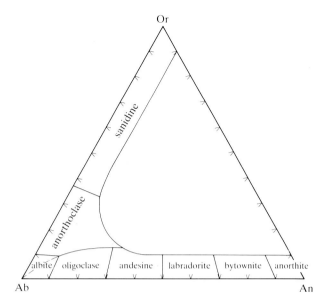

albite molecule and 0% to 10% of the anorthite molecule. This can be written in a more compact way as $Ab_{100}An_0$ to $Ab_{90}An_{10}$, or even more briefly as An_0–An_{10}. Thus the six members of the plagioclase feldspar series have the compositions **albite** An_0–An_{10}, **oligoclase** An_{10}–An_{30}, **andesine** An_{30}–An_{50}, **labradorite** An_{50}–An_{70}, **bytownite** An_{70}–An_{90}, and **anorthite** An_{90}–An_{100}.

The optical properties and structure of the feldspars depend upon their temperature of crystallization and their cooling history. Thus, in quickly cooled extrusive igneous rocks, for example, the potassium feldspar that forms has a tabular crystal habit and is called sanidine, with optical properties peculiar to that mineral. However, when plutonic igneous rocks crystallize, their cooling rate is much slower, which controls the kinetics of order–disorder in the feldspar structure, leading to the formation of orthoclase, with a prismatic habit and slightly different optical properties from those of sanidine.

Much work has been carried out on the feldspar minerals in recent years, and Smith (1974) summarized the differences between the various feldspar types. The composition and nomenclature of the feldspars at various temperatures are illustrated in Figures 2.12–14. The term "temperatures" refers to the temperatures at which effective structural re-equilibrium ceases, and the mineral structure will not be subject to any change thereafter. It is a time–temperature–kinetic relationship. The feldspars are divided into three groups: high (representing extrusive rocks), intermediate

79

Figure 2.12
The high-
temperature
feldspars.

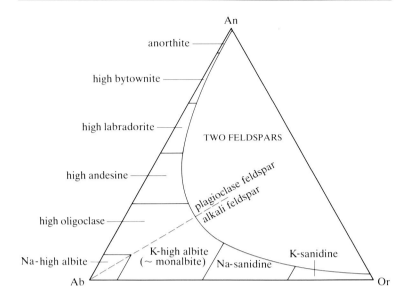

(representing hypabyssal rocks and small intrusions) and low (deep-seated plutonic and metamorphic rocks).

Feldspars quenched from high temperature are shown in Figure 2.12. An arbitrary boundary from albite (Ab) with anorthite (An) equal to orthoclase (Or) defines the alkali feldspar and plagioclase feldspar fields. The plagioclase feldspars are divided into six divisions at 10, 30, 50, 70 and 90 mol % An, and the feldspars in these divisions are named on the diagram as (high) albite 0–10% An, oligoclase 10–30% An, andesine, labradorite, bytownite and anorthite 90–100% An. In the alkali feldspars, the boundary at 40% Or between high albite and high Na-sanidine represents a structural change from triclinic to monoclinic. The boundary at 70% Or, separating Na-sanidine from K-sanidine, is an arbitrary one.

In Figure 2.13 the feldspars formed after short cooling histories (fast rate of cooling) show unmixing with perthite development. Homogeneous feldspars (similar to those found in Fig. 2.12) occupy small areas in Figure 2.13, with K-feldspar (or orthoclase) restricted to a small field. Most alkali feldspars are **perthites**, consisting predominantly of an alkali feldspar host with an exsolved plagio-clase feldspar phase resulting from segregation from an original higher-temperature homogeneous feldspar. Perthites seen with the naked eye are called **macroperthites**; those seen under the micro-scope are called **microperthites**; and those too fine to be distin-guished except by X-ray techniques are called **cryptoperthites**. Cryp-toperthites tend to give the host mineral optical properties between

80

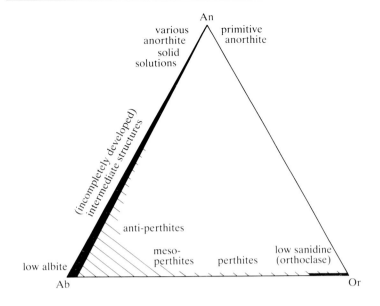

Figure 2.13
Feldspars
formed after
fast cooling

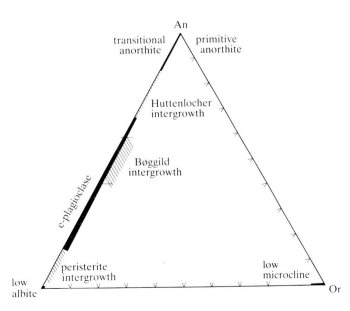

Figure 2.14
Feldspars
formed after slow
(prolonged) cooling

it (the host) and the exsolving phase. When plagioclase feldspar is the host mineral, the unmixed feldspar is called an **antiperthite**. Some perthites consist of roughly equal amounts of intergrown alkali feldspar and plagioclase, and these are called **mesoperthites** (Fig. 2.13). Submicroscopic cryptoperthites may occur, for example in orthoclase, which tend to give the host mineral intermediate optical properties, between those of the host and the exsolved phase.

The feldspar fields are shown after prolonged cooling histories, such as will occur in large plutons or in metamorphic rocks, in Figure 2.14. K-feldspar is restricted to low microcline, and homogeneous plagioclases are restricted to nearly pure albite, anorthite An_{85-100} and the compositional range from An_{15} to An_{70}, with all of these having no more than about 2 mol % Or in the structure. The non-homogeneous types consist of complex series of intergrowths, of which three are important: *peristerites*, containing equal amounts of alkali feldspar and plagioclase; *Bøggild intergrowths*, from roughly An_{40} to An_{60}; and *Huttenlocher intergrowths*, from An_{70} to An_{85}.

The Bøggild intergrowths can contain up to 6% Or and are distinguished by labradorite iridescence (caused by the structure).

Although the individual feldspar types will be described in detail, the general optical properties for feldspars are given below.

COLOUR   Colourless, with occasional white or pale brown patches where alteration to clay minerals has occurred.

HABIT   Euhedral crystals – tabular, or prismatic with large basal faces – may occur as phenocrysts in some extrusive rocks, but most feldspars are either subhedral (prismatic) or anhedral in most rocks.

CLEAVAGE   All feldspars possess two cleavages {001} and {010}, intersecting nearly at right angles on a (100) section. Several partings may occur.

RELIEF   Low, from just below 1.54 (K-feldspar) to above 1.54 (most plagioclases). Details are given in Figure 2.21.

ALTERATION   All feldspars may alter to clay minerals. The individual descriptions give details.

BIREFRINGENCE   Maximum interference colours are first-order white in Ca-poor plagioclase, and first-order yellow in Ca-rich plagioclase.

INTERFERENCE FIGURES   Variable in sign and size. Usually large, so that a single optic axis figure is often required for examination.

EXTINCTION   Details for alkali feldspar are given in Figure 2.18. Plagioclase feldspars show repeated twinning, and the symmetrical extinction angles measured on the twin plane are used to obtain plagioclase composition. These extinction angles are shown in Figures 2.24 and 2.25, but the relevant section in the plagioclase feldspar descriptions must be consulted for details of this technique.

TWINNING   K-rich alkali feldspars exhibit simple twinning, but the plagioclase feldspars show polysynthetic twinning, repeated twinning or complex multiple twins.

ZONING   Common in most plagioclase feldspars, particularly in phenocrysts in extrusive rocks.

PERTHITES   Feldspars frequently show effects of unmixing or exsolution, resulting in intergrowths – plagioclase feldspars within alkali feldspar host and vice versa.

Alkali feldspars There is a continuous series from K-feldspar ($KAlSi_3O_8$) to albite ($NaAlSi_3O_8$) for high and low alkali feldspars.

Sanidine–high albite series

Ab_0–Ab_{63} sanidine
Ab_{63}–Ab_{90} anorthoclase
Ab_{90}–Ab_{100} high albite

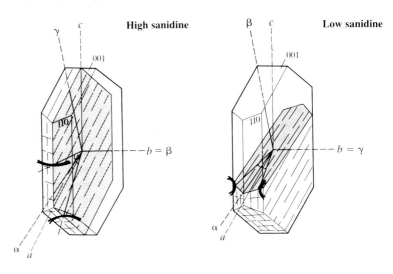

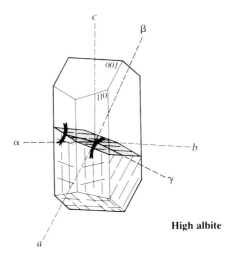

High albite

Orthoclase–low albite series

Or_{100-85} orthoclase
Or_{85-20} orthoclase cryptoperthites
Or_{20-0} low albite

Microcline–low albite series

Or$_{100-92}$ microcline
Or$_{92-20}$ microcline cryptoperthites
Or$_{20-0}$ low albite

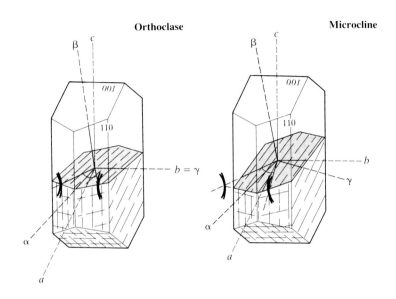

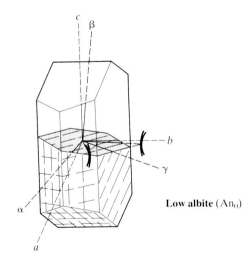

Low albite (An$_0$)

Figure 2.15 A low microcline crystal, showing cross-hatch twinning (×
10, PPL).

There is a continuous variation in RI from K- to Na-alkali feldspar,
and general values for the end members are as follows:

	Ab	Or
n_α	1.527	1.518
n_β	1.531	1.522
n_γ	1.539	1.524
δ	0.012	0.006

$2V$ is variable in size or sign depending on composition and type: the
full range of values is given in Figure 2.16. $2V$ values are 15°–40°
sanidine, 42°–52° anorthoclase and 52°–54° high albite: all are
biaxial negative, as is orthoclase (35°–50°) and microcline (66°–90°).
Low albite is biaxial positive with $2V_\gamma = 84°-78°$
OAP also varies: this variation is given in Figure 2.16 and in the
various feldspar diagrams
$D = 2.56-2.63 \qquad H = 6-6\frac{1}{2}$

COLOUR Colourless with opaque patches if alteration to clay minerals has
occurred.

HABIT Euhedral prismatic in high-temperature porphyritic rocks to anhedral in plutonic intrusive rocks, although some porphyritic granites
contain euhedral (high T) alkali feldspar phenocrysts; for example,
the granite from Shap Fell, Cumbria.

CLEAVAGE Two cleavages {001} and {010} meeting nearly at right angles on the
(100) plane. Several partings may be present.

86

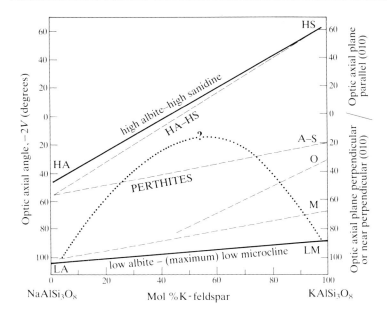

Figure 2.16
2V variation in
alkali feldspars.

RELIEF	Low, less than 1.54.
*ALTERATION	Common, usually to clay minerals with K-feldspar altering as follows in the presence of a limited amount of water:

$$3 \text{ Or} + 2 \text{ H}_2\text{O} \rightarrow \text{illite} + 6 \text{ silica} + 2 \text{ potash}$$

or, if excess water is present,

$$2 \text{ Or} + 3 \text{ H}_2\text{O} \rightarrow \text{kaolin} + 4 \text{ silica} + 2 \text{ potash}$$

The clay minerals found occur as discrete tiny particles held within the feldspar crystal. As the amount of alteration increases, the clays increase in both amount and size, to a point at which they are usually termed sericite (Fig. 2.17). The sodium-rich alkali feldspars can alter in the same way that plagioclase feldspar does, with the clay mineral montmorillonite being formed:

$$\text{Na-feldspar} + \text{H}_2\text{O} \rightarrow \text{montmorillonite} + \text{Qz} + \text{soda}$$

BIREFRINGENCE	Low, first-order grey or sometimes white are maximum colours.
INTERFERENCE FIGURE	2V is usually 40°–65° and negative, except for low albite which has a very large positive 2V. A single optic axis figure will be best, and for this an isotropic section is needed.
EXTINCTION	The extinction angle measured to the {010} cleavage trace varies depending on composition, and is given in Figure 2.18.
*TWINNING	Simple twins are common in the K-rich alkali feldspars, and the common monoclinic twin forms are shown in Figure 2.19, together with the two common triclinic twins. Low microcline, which is the

Figure 2.17 A low-relief, colourless K-feldspar crystal, with higher-relief minerals also present. The K-feldspar shows alteration to cloudy masses of fine-grained sericite. Note how the feldspar cleavages are emphasized by the alteration ($\times$ 10, PPL).

common alklai feldspar type in sedimentary sandstones, meta-morphic rocks and large plutonic acid intrusions, possesses a distinctive cross-hatched type of twinning in which both pericline and albite twin laws operate.

*PERTHITES These intergrowths of a Na-plagioclase in a K-feldspar host are always found in low-temperature alkali feldspars.

*DISTINGUISH- (a) *Sanidine–high albite*. $2V$ angle and extinction angles are small.
ING FEATURES Anorthoclase shows two sets of polysynthetic twins yielding a grid, or cross hatching, similar to microcline. Anorthoclase is confined to extrusive rocks, whereas microcline is found only in plutonic rocks. Microcline has a large $2V$ ($\sim 67°$) and is negative, but an interference figure is virtually impossible to obtain.

(b) *Orthoclase* is difficult to identify and can easily be mistaken for quartz, but its RIS are less than 1.54. Also, it is frequently cloudy from alteration to clays, compared with quartz which is *always* unaltered and clear. Orthoclase is biaxial with a negative $2V_a$, which may increase in size if submicroscopic cryptoperthites exist, thus giving $2V_a$ values between orthoclase (max. 50°) and low albite ($\sim 90°$). Orthoclase is biaxial negative, whereas nepheline is uniaxial negative, and orthoclase has a slightly larger $2V$ than its high-temperature form sanidine.

88

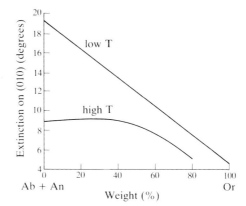

Figure 2.18
Feldspar
extinction
angles.

The alkali feldspars are essential constituents of alkali and acid igneous rocks, particularly syenites, granites and granodiorites, felsites and orthoclase porphyries and trachytes, rhyolites and dacites. Alkali feldspars are common in pegmatites, in hydrothermal veins and in high-grade gneisses. Plutonic rocks contain orthoclase, microcline and perthites, whereas extrusive rocks contain sanidine and other "high" types. Perthite types can be correlated, on the basis of the size of the exsolved phase, with decreasing temperature of formation and length of time of cooling period, from cryptoperthites (< 5 nm obtained by X-ray diffraction data) found in hypabyssal rocks, to macroperthites (> 0.1 mm), which occur in igneous rocks formed from large plutonic intrusions.

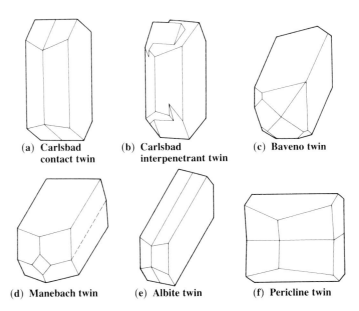

(a) **Carlsbad contact twin**

(b) **Carlsbad interpenetrant twin**

(c) **Baveno twin**

(d) **Manebach twin**

(e) **Albite twin**

(f) **Pericline twin**

Figure 2.19
Feldspar twin
forms.

Figure 2.20 The myrmekitic intergrowth of quartz and feldspar. On a larger scale, in graphic granite, the quartz crystals were once said to resemble writing; hence the name ($\times$ 2.5, XPOLS).

Some pegmatites contain intergrowths of alkali feldspar and quartz called graphic intergrowth (because the quartz crystals resemble writing), and may be due either to simultaneous crystallization of alkali feldspar and quartz or to the replacement of some of the alkali feldspar by quartz (see Fig. 2.20). Associations of feldspar, usually plagioclase but occasionally alkali feldspar, and quartz in which the feldspar and quartz are intercalated in "stringer"-like textures, are called myrmekite. In Rapakivi granites, large crystals of alkali feldspar are mantled by plagioclase, and similar relationships occur in other orbicular granites. Potassium feldspars are common in high-grade metamorphic rocks. It is a stable mineral at the highest grades when the breakdown of muscovite leads to the presence of K-feldspar, with sillimanite also being formed:

$$\text{muscovite} + Qz \rightarrow \text{K-feldspar} + \text{sillimanite}$$

These high-grade metamorphic rocks containing K-feldspar include charnockites, sillimanite gneisses, migmatites and granulites.

Orthoclase and microcline occur as detrital grains in terrigeneous arenaceous rocks: authigenic alkali feldspars, forming at low temperatures within sediments, have been recognized which are usually homogeneous (non-perthitic) and untwinned.

90

Plagioclase feldspars

End members   **high/low albite (Ab)**   NaAlSi$_3$O$_8$ all triclinic
                 high/low anorthite (An)   CaAl$_2$Si$_2$O$_8$ $0.820 \pm : 1.295 \pm : 0.720 \pm$
                                          $a = 93°30' \pm , \beta = 116° \pm , \gamma = 88°–91°$

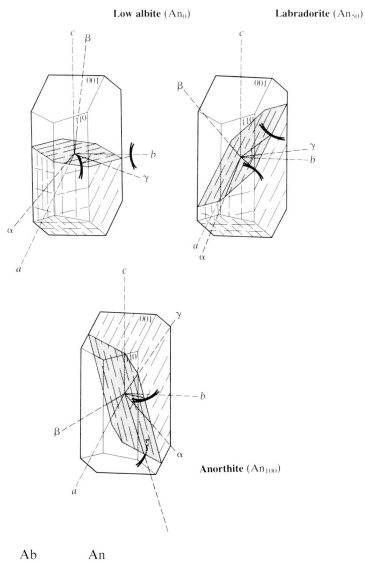

Low albite (An$_0$)   **Labradorite** (An$_{50}$)

Anorthite (An$_{100}$)

	Ab	An
n_α	1.527	1.577
n_β	1.531	1.585
n_γ	1.539	1.590
δ	0.012	0.013

91

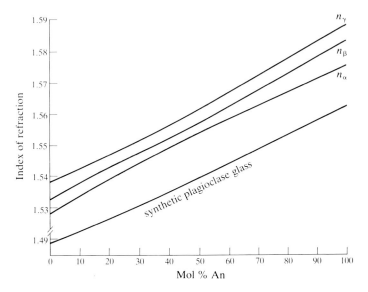

Figure 2.21 RI variation in plagioclase feldspars and glass made from these fledspars

There is continuous variation in the principal RIs of plagioclase feldspars from Ab to An, and this is given in Figure 2.21.

$2V$ is variable depending on the structural state and the composition, and Figure 2.22 shows the curves for variation of $2V$ for high (solid line) and low (dashed line) plagioclase feldspar

$D = 2.62–2.76$ $H = 6–6\frac{1}{2}$

COLOUR	Colourless, occasionally near opaque because of clay development.
HABIT	Subhedral prismatic in plutonic and hypabyssal rocks, to euhedral prismatic or tabular in extrusive rocks.
CLEAVAGE	Similar to alkali feldspars. Two perfect cleavages {001} and {010} meeting at nearly right angles on the (100) plane. Several partings may occur.
RELIEF	Low, but apart from albite all plagioclase feldspars have RIs greater than 1.54.
*ALTERATION	Already partly dealt with under alkali feldspars, but it is worth repeating here that plagioclase feldspar alters either to montmorillonite if limited water is available, or to kaolin if excess water is available. This alteration may be the result of either late-stage hydrothermal activity during solidification of the rock mass or of chemical weathering. Other minerals which may form from feldspars during late-stage hydrothermal activity include the epidote mineral zoisite, or clinozoisite, which is produced during saussuritization (see also the epidote group minerals).

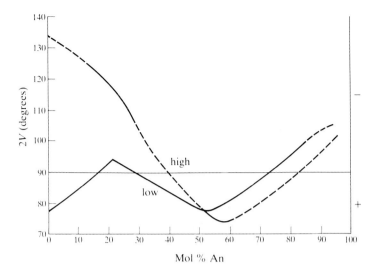

Figure 2.22
2V variation in plagioclase feldspars

BIREFRINGENCE
Low, with interference colours varying from first-order greys (Ab) to first-order yellows (An).

INTERFERENCE
FIGURE
2V is generally large and variable in sign, with the data for high and low series shown in Figure 2.22. The OAP orientation varies from albite to anorthite, and Philips & Griffin (1981) should be consulted for details.

*TWINNING
Multiple twinning by the *albite law* is common in all plagioclase feldspars and is a *characteristic* feature. Albite twin lamellae, which tend to be parallel to the prism zone, often tend to be narrow in the Na-plagioclases, and alternating narrow and broad in the Ca-plagioclases (Fig. 2.23). Other twin laws which operate in plagioclases include *Carlsbad* (simple) and *pericline* (repeated): combinations of twins, such as *Carlsbad–albite*, are common. A combined Carlsbad–albite twin, showing symmetrical extinction in each half of the twin, is shown in Figure 2.24. Sedimentary authigenic plagioclase feldspars (albites) and sodic plagioclases in low-grade metamorphic rocks may be untwinned.

*EXTINCTION
The composition of plagioclases may be determined by measuring the symmetrical extinction angles of albite twins, on sections at right-angles to the *a* crystallographic axis. Full details of the variation in the maximum extinction angle with composition, for both high (extrusive rocks) and low (hypabyssal and plutonic rocks) plagioclase feldspars, are given in Figure 2.25. Combined Carlsbad–albite twins may also be used for determination of composition, but these combined twins are usually only found in ultrabasic igneous plutonic rocks (troctolites and peridotites) and some very basic extrusive types. The curves needed to obtain the composition of a combined twin in plutonic rocks, with each half of the Carlsbad

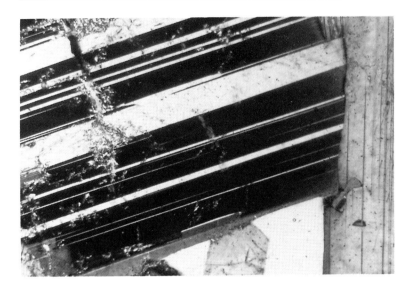

Figure 2.23 Typical albite twinning (also called lamellar or polysynthetic twinning) in a crystal of plagioclase feldspar (× 10, XPOLS).

twin being examined separately, are shown in Figure 2.26. Thus the smaller symmetrical extinction angle of one half of the twin is plotted along the ordinate, and the larger symmetrical extinction angle of the other half of the twin is plotted onto the curves. The composition is then read off along the abscissa. No other twin types are commonly used to determine composition.

*ZONING Zoning is common in plagioclase feldspars from extrusive rocks, normally showing up as a continuous change in composition from a calcium-rich core to a sodium-rich margin. In this case the composition should be given as, for example, An_{70} (core) to An_{32} (margin). If the zoning is reversed (sodium-rich core to calcium-rich margin), the precise variation in the composition should again be given; or if the zoning is oscillatory, where separated zones of equal extinction occur, an indication of the zonal variation should be given.

OCCURRENCE Plagioclase feldspars are almost always present in igneous rocks (with the exception of some ultramafic and ultra-alkaline types) and often comprise more than half the rock's total volume. Plagioclase varies in composition with the type of rock it is found in; thus bytownite occurs in ultrabasic rocks and labradorite in basic rocks, andesine is typical of intermediate rocks, and oligoclase is common in acid rocks.

In basic lavas, calcium-rich plagioclase feldspars occur as both phenocrysts and constituents of the groundmass. In basic plutonic

94

(a) Albite twin

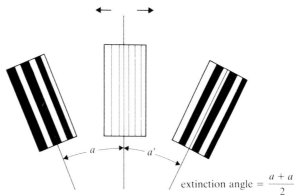

$$\text{extinction angle} = \frac{a + a'}{2}$$

(b) Combined Carlsbad–albite twin

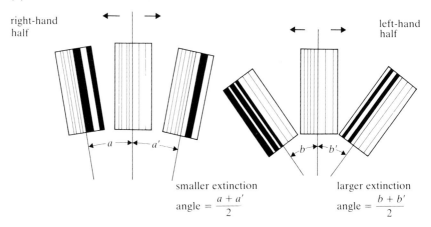

right-hand half

left-hand half

smaller extinction angle $= \dfrac{a + a'}{2}$

larger extinction angle $= \dfrac{b + b'}{2}$

Figure 2.24   The measurement of extinction angles in (a) an albite twin, and (b) a combined Carlsbad–albite twin.

intrusions, layering and differentiation can occur, with feldspar-rich layers being common. In these intrusions plagioclase may show a compositional range from An_{85} to An_{30} and is frequently zoned. The "low" plagioclase found in plutonic rocks is often antiperthite, particularly in acid types, with exsolved alkali feldspar (K-feldspar). In other plutonic rocks, especially those with a long cooling history, peristerites (Fig. 2.14) may occur in which Schiller effects can be seen. The most basic intrusion may show either Bøggild inter-growths, caused by unmixing of two plagioclase components, or Huttenlocher intergrowths, which occur in bytownites in which two basic plagioclase components unmix; however, both of these inter-growth types are rare and are seldom seen optically. Anorthosites contain plagioclase feldspars as the primary constituent, comprising

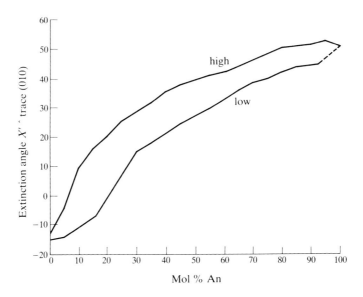

Figure 2.25 The maximum extinction angles for albite twins in high- and low-plagioclase feldspars. Note that measurement is always to the fast ray.

well over 80% of the volume of the rock. The plagioclase composition varies from bytownite to andesine, although with any particular anorthosite intrusion the compositional range is quite small.

Pure albite is the typical feldspar of spilites, often with relict cores of a more anorthitic plagioclase. This may indicate a late-stage magmatic or metasomatic process by which the original feldspar in the basalt becomes increasingly sodium rich, a process called albitization. However, in some spilites the albite may be a primary crystallizing mineral.

In metamorphic rocks, the composition of the plagioclase reflects the metamorphic grade of the rock, the plagioclase becoming more calcium rich as the grade increases. Albite is the typical plagioclase of low-grade regional rocks, with oligoclase occurring at garnet grade. In granulites and charnockites, andesine or, rarely, labradorite is the common plagioclase. Plagioclase feldspars do not occur in eclogites, the various feldspar components (Ca, Al, etc.) entering either the clinopyroxene or garnet phases. Pure anorthite may occur in thermally metamorphosed calcareous rocks.

In sedimentary rocks, apart from plagioclase occurring as detrital grains in many terrigeneous arenaceous rocks, albite may occur as an authigenic mineral in some sandstones, forming during sedimentation. Authigenic albite may be simply twinned, but never shows lamellar twinning.

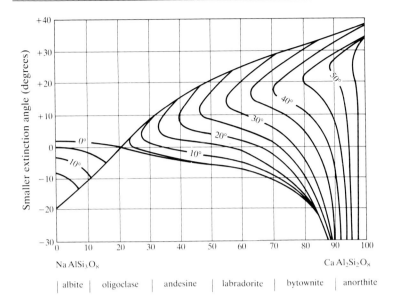

Figure 2.26 The extinction angles of Carlsbad–albite twins. The extinction angle for an albite twin is measured in each half of a Carlsbad twin. The smaller angle is plotted along the ordinate and the larger angle *into* the nest of curves (see Fig. 2.24 for details of measurement of albite twin extinction angles). Thus, for example, a Carlsbad–albite twin with a refractive index of greater than 1.54, and with angles of extinction of 10° (the smaller) and 30° (the larger) has a composition of An_{60}. The negative ordinate values (below the horizontal line representing 0°) are needed for feldspars which have refractive indices of less than 1.54.

In addition to the normal feldspar group of minerals already discussed, some feldspars containing more than 2% BaO are termed barium feldspars. The most common barium feldspar ($BaAl_2Si_2O_8$), **celsian**, is similar in almost every property to the feldspars, especially orthoclase, except that it has greater RIS and a much higher density. It is rare and tends to be found associated with manganese deposits and stratiform barite deposits. The mineral **hyalophane** $(K,Na,Ba)[(Si,Al)_4O_8]$, with K > Ba, is another barium-rich feldspar, similar in optical properties to celsian and with similar modes of occurrence.

Feldspathoid family   Tektosilicates
The group of minerals termed "feldspathoids" include those minerals which have certain similarities with the feldspars, particularly in their chemistry and structure. The main feldspathoid minerals given in detail here are:

leucite $KAlSi_2O_6$
nepheline $NaAlSiO_4$
sodalite $Na_8Al_6Si_6O_{24}Cl_2$ or $6(NaAlSiO_4).2NaCl$

Other feldspathoid minerals include:

hauyne $CaNa_6Al_6Si_6O_{24}.So_4$ or $6(NaAlSiO_4).CaSO_4$, with S
             replacing SO_4
nosean $Na_8Al_6Si_6O_{24}SO_4$ or $6(NaAlSiO_4).Na_2SO_4$
cancrinite   complex hydrated alkali aluminosilicate with CO_3,
             SO_4 and Cl groups included
kalsilite $KAlSiO_4$

The feldspathoids are all silica deficient compared with the feldspars, and their occurrence is restricted to undersaturated alkali igneous rocks. Although analcime is a zeolite it is very closely associated with the feldspathoids, and has been included here after sodalite.

Leucite $KAlSi_2O_6$ tetragonal
                                        (pseudo-cubic) c/a 1.054

$n = 1.508–1.511$
$\delta = 0.001$
If uniaxial, leucite is +ve but the sign is virtually impossible to determine because of twinning
$D = 2.47–2.50$ $H = 5\frac{1}{2}–6$

COLOUR   Colourless.
*HABIT   Usually euhedral crystals showing eight-sided sections.
CLEAVAGE   Very poor on {110}.
RELIEF   Low.
BIREFRINGENCE Isotropic to very low.
*TWINNING   Repeated twinning on {110} is always present and visible under crossed polars, as a type of cross hatching (Fig. 2.27).
OTHER   At high temperatures, leucite can contain sodium in the structure,
FEATURES   which must be expelled at lower temperatures. Pseudo-leucite, a mixture of nepheline and feldspar, forms an intergrowth which completely replaces leucite in some rocks.
*OCCURRENCE In potassium-rich basic extrusive rocks such as leucite–basanite, leucite–tephrite and leucitophyre, which are usually silica deficient. Pseudo-leucite may occur in some alkali basic plutonic rocks, but mainly occurs in extrusive igneous rocks. At subsolidus temperatures, such as can be attained in plutonic intrusions, leucite breaks down to give nepheline and feldspar, which explains the absence of leucite and the presence of nepheline and feldspar assemblages in plutonic alkali rocks.

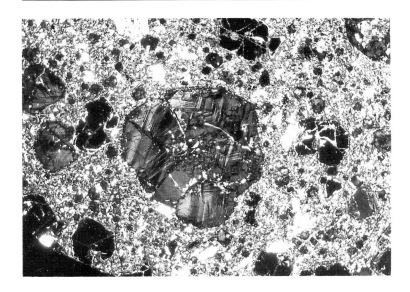

Figure 2.27 Leucite crystals in a basic undersaturated lava flow. The large central crystal exhibits complex cross hatch twinning, and is weakly isotropic (× 1.25, XPOLS).

Nepheline NaAlSiO$_4$ hexagonal
(K may replace Na up to 25%) c/a 0.838

n_o = 1.529–1.546
n_e = 1.526–1.542
δ = 0.003–0.005
Uniaxial − ve (crystal is length fast)
D = 2.56–2.66 $H = 5\frac{1}{2}$–6

COLOUR   Colourless.
HABIT   Usually anhedral, occurring in the interstices between minerals. Occasionally found as small exsolved 'blebs' within feldspars, particularly K-feldspars. Euhedral crystals have an hexagonal outline.
CLEAVAGE   {10$\bar{1}$0} imperfect prismatic cleavage, and poor basal {0001} cleavage.
RELIEF   Low.
ALTERATION Nepheline may alter to zeolites, such as natrolite Na$_2$Al$_2$Si$_3$O$_{10}$.2H$_2$O or analcime, and to feldspathoids such as sodalite, e.g.

$$2NaAlSiO_4 + SiO_2 + 2H_2O \rightarrow Na_2Al_2Si_3O_{10}.2H_2O \quad \text{(natrolite)}$$
$$NaAlSiO_4 + SiO_2 + H_2O \rightarrow NaAlSi_2O_6.H_2O \quad \text{(analcime)}$$

by the addition of silica, water and other volatiles (chlorine in the case of sodalite). Nepheline commonly alters to cancrinite (a complex hydrated silicate with sulphate, carbonate and chloride).

99

*BIREFRINGENCE   Low, first-order greys. Small inclusions occur within nepheline, and give the crystal a "night sky" effect under crossed polars.

TWINNING   Rare.

*OCCURRENCE   A characteristic primary crystallizing mineral of alkali igneous rocks. Nepheline is an essential constituent of silica-deficient nepheline–syenites, and may occur in volcanic rocks where it is associated with high temperature feldspars.

Nepheline may be metasomatic in origin, formed by the reaction of alkali-rich magmatic fluids with country rocks. Nepheline may occur in basic rocks near their contact with carbonate rocks; and alkali dolerites with interstitial nepheline have also been described.

Another mineral associated with nepheline is **kalsilite** ($KAlSiO_4$) which forms a limited solid solution with $NaAlSiO_4$. Up to 25% of the nepheline molecule can be replaced with kalsilite, although the amount of solid solution increases with increasing temperature, up to a maximum of 70% at 1,070°C. Kalsilite is not found in plutonic igneous rocks but is found in the groundmass of some K-rich lavas.

Sodalite $Na_8Al_6Si_6O_{24}Cl_2$ cubic

$n = 1.483–1.487$
$D = 2.27–2.88$ $H = 5\frac{1}{2}–6$

COLOUR   Colourless, sometimes pale blue or pink.

HABIT   Usually anhedral, but occasional eight-sided anhedral crystals occur.

CLEAVAGE   Poor {110}.

RELIEF   Moderate (much less than CB).

*DISTINGUISHING   Sodalite and analcime are virtually indistinguishable in thin section.
PROPERTIES   The type of rock is as good a guide as any as to which mineral it is.

OCCURRENCE   Found in nepheline–syenites in association with nepheline and fluorite. It occurs in metasomatized calcareous rocks near alkaline intrusions. **Lapis lazuli** is a member of the sodalite group of minerals, being dark blue in colour, and occurring in contact altered limestones.

Nosean   Two related minerals, nosean ($Na_8Al_6Si_6O_{24}SO_4$) and hauyne,
Hauyne   which has a similar composition to nosean with some S and Ca in
Cancrinite   the structure, have virtually identical optical properties. Nosean is distinguished by having a dark border around each crystal, and it usually occurs in silica-poor alkaline rocks such as phonolites and leucitophyres. Hauyne is indistinguishable from sodalite and it also occurs in phonolites and other undersaturated rocks, but is usually accompanied by sulphide minerals such as pyrite. Cancrinite shows higher birefringence than nepheline (first order yellows and reds), is

uniaxial negative, and is a characteristic mineral in rocks similar to those in which nepheline occurs. Cancrinite may be an alteration product of nepheline.

Analcime (Analcite)

$NaAlSi_2O_6.H_2O$ cubic

$n = 1.479–1.493$

$D = 2.34–3.29$ $H = 5\frac{1}{2}$

COLOUR	Colourless.
HABIT	Anhedral crystals filling interstices between mineral grains.
CLEAVAGE	Poor {001} fracture.
RELIEF	Moderate (less than 1.54).
*OCCURRENCE	It has been suggested that the analcime in some intermediate and basic rocks such as dolerite, teschenites and essexites (porphyritic alkali gabros) is primary, but recent work has questioned this. In volcanic rocks analcime occurs in some basalts as a primary mineral, but most commonly as a late-stage hydrothermal mineral crystallizing in vesicles and found with zeolites – especially thomsonite and stilbite.

Analcime occurs as an authigenic mineral in sandstones, again associated with the zeolites laumontite and heulandite. Analcime has also been found in pyroclastic rocks.

Garnet group   Nesosilicates

$X_3Y_2Si_3O_{12}$ cubic

			n	D	
Almandine	(X = Fe)	(Y = Al)	1.830	4.318	
Pyrope	(X = Mg)	(Y = Al)	1.714	3.582	
Grossular	(X = Ca)	(Y = Al)	1.734	3.594	$H = 6–7\frac{1}{2}$
Spessartine	(X = Mn)	(Y = Al)	1.800	4.190	
Andradite	(X = Ca)	(Y = Fe)	1.887	3.859	

Al^{3+} occupies the Y position in the above minerals, but Cr^{3+}, Fe^{3+} and Ti^{4+}, as well as REES, may also substitute in this position in garnet minerals. Solid solutions covering a wide range are possible within the garnet group of minerals, with a corresponding wide range of optical properties: RI varies from 1.890 to 1.710, and density from 4.32 to 3.59. A variety called **hydrogrossular** $Ca_3Al_2[SiO_4]_{1-m}(OH)_{4m}$ possesses an RI as low as 1.68 and a density as low as 3.1.

COLOUR	Colourless, pale brown or pale pink, dark green or brown.
*HABIT	Euhedral crystals of garnet showing six-sided or eight-sided sections are common.

101

*CLEAVAGE	No cleavage exists, but crystals are often fractured.
*RELIEF	High to very high; sections always appear to have a rough surface (see Plate 1a).
ALTERATION	Fe–Mg-bearing garnets typically alter to chlorite (by hydrothermal reaction), with quartz being produced in the reaction. Garnet is isotropic (see Plate 1b) but sometimes hydrogrossular in particular may be anisotropic and zoned.
*OTHER FEATURES	Garnet occasionally exhibits compositional zoning, and in meta-morphic medium-grade rocks garnet frequently contains inclusions of quartz and micas. Some metamorphic garnets contain inclusions defining an earlier fabric (e.g. schistosity) which has been trapped within the garnet.
OCCURRENCE	Garnet occurs mainly in metamorphic rocks. Almandine (contain-ing some Mg) is the main garnet of middle-grade metamorphism of pelitic rocks and igneous rocks. Almandine forms by progressive metamorphic reactions involving chlorite, and by a reaction between quartz and staurolite which gives garnet and kyanite. In high heat flow, low-P metamorphism garnet may form from cordi-erite breakdown if the chemistry is correct. Grossular forms in thermal aureoles in, and regional metamorphism of, impure lime-stones; and spessartine forms during the metamorphism of Mn-rich rocks. Andradite occurs in thermally metamorphosed impure cal-careous sediments, and particularly in metasomatic skarns. Hydro-grossular is found in metamorphosed marls, altered gabbros and rodingites, which are altered calc-silicate rocks associated with serpentinites. The black variety **melanite** is found in alkali igneous rocks.

Pyrope garnet is an essential constituent of some ultrabasic igneous rocks, especially garnet–peridotite and other similar types derived from the Earth's upper mantle. High-grade metamorphic rocks (very high P and T) called eclogites have garnet of almandine–pyrope composition as an essential constituent, along with the pyroxene omphacite.

Garnet also occurs as a detrital mineral in sands.

Humite group
$n\text{Mg}_2[\text{SiO}_4]\text{Mg(OH,F)}_2$

Nesosilicates
orthorhombic
$0.463:1:\dfrac{0.855 \text{ (No)}}{2.057 \text{ (Hu)}}$

There are four members of the group; nor-bergite (with $n = 1$), chondrodite ($n = 2$), humite ($n = 3$) and clinohumite ($n = 4$).

monoclinic
$2.170:1:1.663$ (Ch)
$0.462:1:1.332$ (Cl)

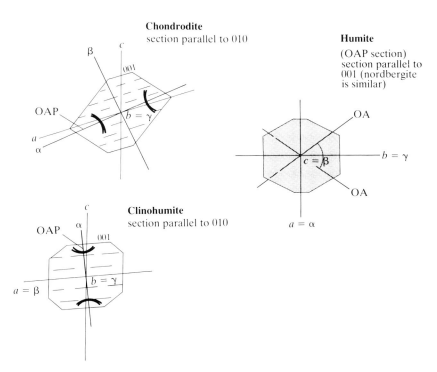

Chondrodite
section parallel to 010

Humite
(OAP section)
section parallel to
001 (nordbergite
is similar)

Clinohumite
section parallel to 010

	n_α	n_β	n_γ	δ
Norbergite	1.561–1.567	1.567–1.579	1.587–1.593	0.026
Chondrodite	1.592–1.643	1.602–1.655	1.619–1.675	0.025–0.037
Humite	1.607–1.643	1.619–1.653	1.639–1.675	0.028–0.036
Clinohumite	1.623–1.702	1.636–1.709	1.651–1.728	0.028–0.045

Variation in the RI is caused by Fe^{2+} and Ti^{4+} entering the structure.

	$2V_\gamma$	D	
Norbergite	44°–50° + ve	3.15–3.18	} $H = 6\frac{1}{2}$
Chondrodite	64°–90° + ve	3.16–3.26	
Humite	65°–84° + ve	3.20–3.32	} $H = 6$
Clinohumite	52°–90° + ve	3.21–3.35	

103

*COLOUR	Pale yellow or yellow.
PLEOCHROISM	Norbergite, chondrodite and humite have α pale yellow, β colourless or rarely pale yellow, and γ colourless. Clinohumite has α golden yellow, and β and γ pale yellow.
HABIT	Anhedral masses of crystals usually occur. Occasionally large subhedral porphyroblasts can be present.
CLEAVAGE	Basal {001} usually present.
RELIEF	Moderate to high (clinohumite).
ALTERATION	All the humite minerals alter to serpentine or chlorite, as follows:

$$Mg_2SiO_4Mg(OH,F)_2 + H_2O + SiO_2 \rightarrow Mg_3Si_2O_5(OH,F)_4$$

*BIREFRINGENCE	Moderate to high (clinohumite) giving maximum upper second-order interference colours (lower third-order, clinohumite).
INTERFERENCE FIGURE	The single optic axis figure yields a positive $2V$ which varies in size in different humite minerals (see above).
EXTINCTION	Norbergite and humite have straight extinction on cleavage (both orthorhombic), whereas chondrodite has $\alpha\hat{\ }cl = 3°–12°$ and clinohumite $\alpha\hat{\ }cl = 0°–4°$ (both monoclinic forms).
TWINNING	Simple or multiple twinning on {001} in monoclinic forms.
ZONING	Common, shown by colour intensities.
DISTINGUISH-ING FEATURES	Similar to olivine except for a pale yellow colour with moderate interference colours. The yellow colour in olivine normally implies alteration to serpentine with low or anomalous interference colours. The olivine $2V$ is very large and usually negative. Staurolite has higher RIs, lower birefringence and occurs in schists. Individual humite group members are difficult to distinguish from each other.
OCCURRENCE	The humite minerals have a restricted occurrence, being found in contact metamorphosed and metasomatized limestones and dolomites, near acid or alkaline intrusions.

Lawsonite   Sorosilicates

Lawsonite $CaAl_2[Si_2O_7](OH)_2.H_2O$ orthorhombic
                                                                1.545:1:2.314

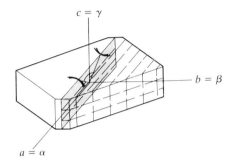

$n_\alpha = 1.665$
$n_\beta = 1.672$
$n_\gamma = 1.684$
$\delta = 0.019$
$2V_\gamma = 76°–87°$ + ve
OAP is parallel to (010)
$D = 3.1$ $H = 7–8$

COLOUR — Colourless in sections of normal thickness; thicker sections may be yellowish.

PLEOCHROISM — Thick sections are pleochroic in yellows and blues, with α pale blue or pale yellowish brown, β pale bluish-green or pale yellow, and γ colourless or pale yellow.

*HABIT — Euhedral crystals are common, and are tabular on {001}. Basal sections are lozenge-shaped.

*CLEAVAGE — Perfect on {001} and {010}, intersecting at right angles in the (100) section. Imperfect prismatic {110} cleavages are present.

RELIEF — Moderate to high.

ALTERATION — Lawsonite is stable at low temperatures, and is an alteration product of plagioclase, to which it returns at high temperatures. It may be replaced by pumpellyite.

BIREFRINGENCE — Low first-order whites and yellows.

INTERFERENCE FIGURES — A large positive $2V$ occurs on a basal section, but a single isogyre is needed to determine the sign.

*EXTINCTION — Straight on cleavages.

DISTINGUISHING FEATURES — Lawsonite is very similar to epidote group minerals, but has a higher birefringence than zoisites, and epidote is negative with oblique extinction. Andalusite and scapolite are negative.

OCCURRENCE | Lawsonite forms from plagioclase by low-temperature meta-morphism or hydrothermal alteration. It is a characteristic mineral of glaucophane schists (**blueschists** – high pressure with low heat flow), where it is associated with pumpellyite, albite, aegirine and jadeite. It may be a secondary mineral, being derived from plagioclase feldspar by hydrothermal alteration of basic igneous rocks.

Melilite group   Sorosilicates
$(Ca,Na)_2(Mg,Al)[(Al,Si)_2O_7]$ tetragonal
                                       c/a 0.65

This group comprises a series from the Ca,Al end-member gehlenite to the Ca,Mg end member åkermanite. The mineral melilite occupies an intermediate position, with Na and Fe^{2+} included as well as Ca, Mg and Al.
$n_0 = 1.670$ (Geh)–1.632 (Åk)
$n_e = 1.658$ (Geh)–1.640 (Åk)
$\delta = 0.012$–0.000 (Geh); 0.000–0.008 (Åk)
Uniaxial +ve (Geh), −ve (Åk)
$D = 3.04$–2.944 $H = 5$–6

COLOUR | Colourless, variable in browns.
PLEOCHROISM | Absent from all except thick sections of melilite with o golden brown and e colourless.
*HABIT | Melilite group minerals usually appear as subhedral to euhedral crystals, tabular on the basal plane.
CLEAVAGE | {001} moderate, {110} poor.
RELIEF | Moderate.
ALTERATION | Melilite alters to fibrous masses of **cebollite** $Ca_5Al_2Si_3O_{14}(OH)_2$ or prehnite $Ca_2Al(AlSi_3)O_{10}(OH)_2$. The latter reaction is as follows:

$$Ca_2Al_2SiO_7 + H_2O + 2SiO_2 \rightarrow Ca_2Al_2Si_3O_{10}(OH)_2$$

Other minerals such as calcite, zeolite and garnet have been reported as alteration products from melilite breakdown.
*BIREFRINGENCE | Low to very low: anomalous interference colours may be seen, with dark blue instead of first-order grey.
INTERFERENCE FIGURE | Difficult to obtain, with poorly defined isogyres on a pale grey field.
ZONING | Oscillatory zoning common.
DISTINGUISHING FEATURES | A very low birefringence and anomalous interference colours are distinctive, as is the limited occurrence.
*OCCURRENCE | Melilite occurs in the groundmass of some silica-poor calcium-rich extrusive basalts (melilitites). Melilite may form under high T, low P metamorphism (sanidinite facies) at the contact between silica-poor magmas and carbonate sediments. Melilite may occur in some slags.

Mica group Phyllosilicates

The micas contain sheets of cations such as Fe, Mg or Al (the octahedral sheets), which are linked to two sheets of linked (SiO_4) tetrahedra: the tetrahedral sheets and the mica minerals belong to the 2:1 layer silicates because of this ratio of tetrahedral to octahedral sheets. The complete "sandwich unit" is linked to another similar unit by weakly bonded K, Na or Ca cations. The chemistry of the micas gives rise to the different micaceous mineral groups, as follows:

Interlayer cations

Normal micas	K
Paragonite micas	Na
Brittle micas	Ca

Cations in tetrahedral coordination

Talcs	Si_8
Phengite micas	Si_7Al
Normal micas	Si_6Al_2
Intermediate micas	Si_5Al_3
Brittle micas	Si_4Al_4

The perfect mica cleavage occurs along the interlayer cations. The general formula for the normal micas is:

$$X_2Y_{4-6}[Si_6Al_2O_{20}](OH,F)_4$$

with $X = K,Na(Ba,Rb,Cs,Ca)$ and $Y = Mg,Fe,Fe^3,Al(Mn,Li,Ce, Ti,V,Zn,Co,Cu)$.

Other micas which may be encountered in rocks include **lepidolite** (Li-bearing, colourless), **zinnwaldite** (K- and ferrous-bearing, pleochroic in pale yellows and grey browns) which may be found in similar granite pegmatites, and **paragonite** (Na-bearing, colourless), found in some sodium-rich metamorphic rocks.

Biotite group This includes a group of minerals which appear to have complete mutual solid solution between four end-members. The end members are:

Phlogopite	K_2Mg_6	$[Si_6Al_2O_{20}](OH,F)_4$
Annite	K_2Fe_6''	$[Si_6Al_2O_{20}](OH,F)_4$
Eastonite	$K_2(Mg_5Al)$	$[Si_6Al_2O_{20}](OH,F)_4$
Siderophyllite	$K_2(Fe_5Al)$	$[Si_6Al_2O_{20}](OH,F)_4$

Biotite $K_2(Mg,Fe'')_6[Si_6Al_2O_{20}](OH,F)_4$ is not one of the four end-members, but the system is arbitrarily divided into **phlogopites**, with Mg:Fe ratios greater than 2:1, and **biotites**, with Mg:Fe ratios less than 2:1. Unusually, iron-rich (ferrous and ferric) biotites are called

lepidomelanes. The optical properties of phlogopite and biotite are given in detail below.

Phlogopite $K_2(Mg,Fe)_6[Si_6Al_2O_{20}](OH,F)_4$ monoclinic

pseudo-hexagonal

$0.576:1:2.206,\ \beta = 99°\,18'$

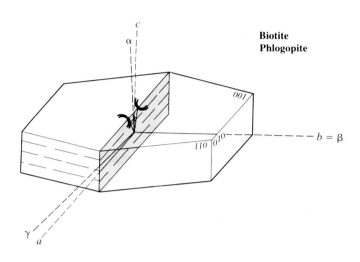

$n_\alpha = 1.530–1.590$ ⎫ RIS increase with increasing iron content,
$n_\beta = 1.557\ 1.637$ ⎪ although the presence of Mn and Ti will also
$n_\gamma = 1.558–1.637$ ⎪ increase RIS. RIS will decrease with increasing F
$\delta\ = 0.028–0.049$ ⎭ content
$2V_a = 0–15°\ -$ ve
OAP is parallel to (010)
$D = 2.76–1.90 \qquad H = 2\frac{1}{2}$

COLOUR   Pale brown, colourless.
*PLEOCHROISM   Weak, with pale colours; a yellow, and β and γ brownish-red, green and deeper yellow.
*HABIT   Small tabular crystals common, frequently subhedral.
*CLEAVAGE   Perfect {001} cleavage.
RELIEF   Low to moderate.
*BIREFRINGENCE High, third-order colours commonly present; weak body colours only slightly mask the interference colours.
*INTERFERENCE FIGURE   A basal section shows a good Bx_a figure with a very small $2V$.
EXTINCTION   Usually straight, but slight angle $\beta\hat{\ }$cleavage = 5° (max) on (010) face may occur.
TWINNING   Rare on {310}, seen on cleavage section.

OTHER FEATURES	Reaction rims may occur in phlogopites found in kimberlite intrusions.
*OCCURRENCE	Found in metamorphosed impure magnesian limestones where phlogopite forms by reactions between the dolomite and either potash feldspar or muscovite. Phlogopite is a common constituent of kimberlite, occurs in many leucite-bearing rocks, and is a minor constituent of ultramafic rocks.

Biotite $K_2(Mg,Fe)_{6-4}(Fe^{3+},Al,Ti)_{0-2}[Si_{6-5}Al_{2-3}O_{20}](OH,F)_4$ monoclinic
(identical to phlogopite)

$n_a = 1.565–1.625$
$n_\beta = 1.605–1.696$
$n_\gamma = 1.605–1.696$
$\delta = 0.040–0.080$
$2V_a = 0–25° - \text{ve}$
OAP is parallel to (010)
$D = 2.7–3.3$ $H = 2\frac{1}{2}$

COLOUR	Brown or yellowish; occasionally green.
*PLEOCHROISM	Common and strong, with a yellow, and β and γ dark brown. Note that pleochroism cannot be detected in a basal section, and that a prism section showing a cleavage is best. Plate 1a shows biotite plates in different orientations.
*HABIT	Tabular, subhedral hexagonal plates.
*CLEAVAGE	{001} perfect.
RELIEF	Moderate.
*ALTERATION	Common in rocks which have undergone hydrothermal alteration. The biotite alters to chlorite, with potash being released in the reaction. The reverse reaction occurs during progressive metamorphism, with chlorite changing in composition as biotite forms at higher temperatures.
*BIREFRINGENCE	High to very high but masked by body colour (see Plate 1b). Note that the birefringence of basal sections of all micas is virtually zero since the n_β and n_γ values are almost equal, and n_β and n_γ lie in the (001) plane (i.e. the basal section).
*INTERFERENCE FIGURE	A basal section gives a perfect Bx_a figure with a very small $2V$, although the strong body colour may tend to mask the colour that determines the sign of the mineral.
EXTINCTION	Nearly straight on cleavage. A speckled effect is seen near the extinction position, which is very characteristic of micas.
TWINNING	Extremely rare, similar to phlogopite twins.
OCCURRENCE	A common mineral in a variety of rocks. Biotite occurs in most metamorphic rocks which have formed from argillaceous sediments. It forms after chlorite and is a constituent of regional metamorphic rocks existing up to very high grades, although its composition

changes with the metamorphic grade. These changes may be accompanied by changes in colour, due to Mg and Ti increasing in amount in the mineral.

Biotites are primary crystallizing minerals in acid and intermediate plutonic igneous rocks, and in some basic rocks. Biotite is not common in acid and intermediate extrusive and hypabyssal rocks.

Biotite is a common mineral in clastic arenaceous sedimentary rocks, but is very prone to oxidation and degradation.

Muscovite $K_2Al_4[Si_6Al_2O_{20}](OH,F)_4$

monoclinic
$0.574:1:2.220, \beta = 95°30'$

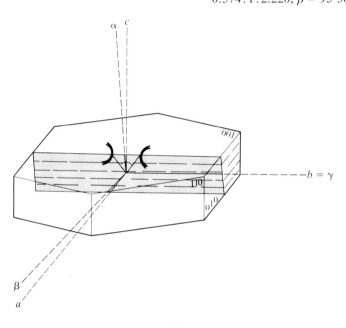

$n_\alpha = 1.552–1.574$
$n_\beta = 1.582–1.610$
$n_\gamma = 1.587–1.616$
$\delta = 0.036–0.049$
$2V_a = 30°–47° - ve$
OAP is perpendicular to (010)
$D = 2.77–2.88$ $H = 2\frac{1}{2} +$

COLOUR   Colourless.

PLEOCHROISM   Muscovite is colourless, but a Cr-rich variety called **fuchsite** is pleochroic in greens.

HABIT   Thin platy crystals; occasionally in aggregates. When the aggregates consist of very fine-grained crystals the mineral is called sericite.

*CLEAVAGE   {001} perfect.

110

RELIEF	Low to moderate (if iron enters the structure).
ALTERATION	Absent.
*BIREFRINGENCE	High, upper second-order to third order (see Plate 1b). Basal sections show a low birefringence of first-order grey which has a characteristic speckled appearance.
*INTERFERENCE FIGURE	A basal section gives an excellent Bx_a figure which just fits into the field of view; i.e. both isogyres will just be seen, indicating a $2V$ size of $40° \pm$.
EXTINCTION	Straight on cleavage.
TWINNING	Similar to phlogopite but not observable.
OCCURRENCE	Muscovite is found in low-grade metamorphic pelitic rocks, where it forms from pyrophyllite or illite. Muscovite remains in these rocks as the grade increases, eventually becoming unstable at highest temperatures ($> 600°C$), when the following reaction occurs:

$$\text{muscovite} + \text{quartz} \rightarrow \text{K-feldspar} + \text{sillimanite}$$

Muscovite occurs in acid igneous plutonic rocks where it is a late crystallizing component, and it occurs in some detrital (clastic) arenaceous sedimentary rocks.

Glauconite $(K,H_3,O)_2(Fe^{3+},Al,Mg,Fe_{2+})_4[Si_{7-7.5}Al_{1-0.5}O_{20}](OH)_4$ monoclinic

$0.577:1:2.208, \beta = 100°$

$n_a = 1.56–1.61$ } RIS increase with increasing ferric iron

$n_\beta = n_\gamma = 1.61–1.65$ } content

$\delta = 0.014–0.032$

$2V_a = 0°–20°$ $-$ ve

OAP is parallel to (010)

$D = 2.4–3.0$ $H = 2$

COLOUR	Pale yellow or green.
*PLEOCHROISM	Glauconite is pleochroic, the colour increasing with increasing iron content, with a pale yellow or yellow-green to green, and β and γ pale to dark green, olive green or bluish green.
*HABIT	Glauconite occurs as rounded pellets or granules with a fine aggregate structure, or as flakes associated with clay minerals.
CLEAVAGE	Perfect on {001}.
RELIEF	Moderate; higher if iron content high.
ALTERATION	Commonly weathered to mixtures of limonite and goethite, and eventually to montmorillonite with the loss of K ions.
BIREFRINGENCE	The maximum birefringence, seen in (010) sections, is lower second-order colours, often dulled by the mineral body colours.
INTERFERENCE FIGURE	Virtually impossible to obtain due to the fine grain size of glauconite, but large flakes are isotropic in basal section with a good small $2V$ seen.
EXTINCTION	Straight on cleavage.

*OCCURRENCE Glauconite occurs as small disseminated diagenetic pellets or granules in detrital sediments of marine origin ("greensands"), chalk marl and chloritic marl. It is associated with clays and chlorites in the pellets, and may be found with carbonates and **collophane** (a cryptocrystalline phosphate mineral) in the marls. Glauconite may be slowly formed by the action of sulphate-reducing bacteria on volcanic glass. It may also be derived from the alteration of biotites.

Olivine group   Nesosilicates

The olivine group consists of an isomorphous series between the two end-members *forsterite* Mg_2SiO_4 and *fayalite* Fe_2SiO_4. Ni and Co may substitute for Mg and Fe in the structure.

Forsterite (Fo)	Mg_2SiO_4	orthorhombic
Fayalite (Fa)	Fe_2SiO_4	$0.467:1:0.587$
		$0.458:1:0.579$

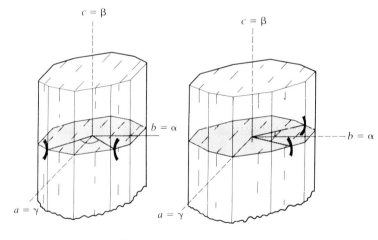

olivine group

$n_\alpha = 1.635$ (forsterite)–1.824 (fayalite)
$n_\beta = 1.651$ –1.864
$n_\gamma = 1.670$ –1.875
$\delta = 0.035$ –0.051
$2V_\alpha = 82°–134°$, i.e. $2V_\gamma = 82°–90°$ + ve, $2V_\alpha = 90°–46°$ – ve
OAP is parallel to (001)
$D = 3.22$ (forsterite) – 4.39 (fayalite) $H = 6\frac{1}{2}$

COLOUR Usually colourless, but may appear pale yellow if the Fe^{2+} content is very high.

112

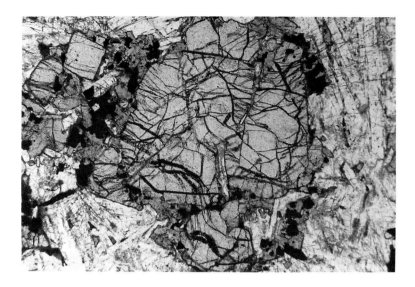

Figure 2.28 Large crystals of olivine in gabbro, exhibiting high relief and irregular fractures which allow ingress of fluids into the crystal structure, causing alteration of the olivine to serpentine and other products ($\times$ 2.5, PPL).

PLEOCHROISM   Extremely rare; only Fa olivines may show α and β pale yellow, with γ yellow.

HABIT   In igneous rocks, olivine ranges from anhedral (plutonic rocks) to subhedral (extrusive rocks). A rough six-sided crystal shape can be seen in basalts, where the olivine occurs as phenocrysts.

CLEAVAGE   {010} poor with rare {100} imperfect fracture (Fig. 2.28).

RELIEF   Variable depending upon composition; forsterite is moderate to high and fayalite very high.

*ALTERATION   Olivine is very susceptible to hydrothermal alteration, low-grade metamorphism and the effects of weathering. Alteration products are usually complex mixtures of fine-grained minerals, which are difficult to recognize by optical means. Three chracteristic forms of olivine alteration can be distinguished, and are given the generalized names **iddingsite**, **chlorophaeite** and **serpentine**. The most common reaction is that of olivine altering to serpentine, as follows:

$$3 \text{ olivine} + 4 \text{ water} + \text{quartz} = \text{serpentine}$$

and serpentine may continue to break down further.

Some alteration of olivine is usually present along irregular basal fractures and poor cleavages, and complete alteration can occur, with the olivine crystal totally replaced by serpentine, often with a release of iron ores.

113

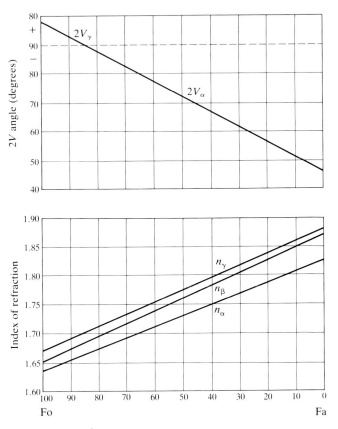

Figure 2.29 Variation of the $2V$ angle and indices of refraction for the forsterite–fayalite (olivine) series (after Bowen & Schairer 1935).

*BIREFRINGENCE	High, with maximum interference colour lower third order (for iron-rich olivines).
*INTERFERENCE FIGURE	$2V$ is generally very large, and a single isogyre should be examined from an isotropic section. Such a grain will not appear black on rotation but will appear greyish, owing to the high relief and dispersion of olivine. The optic axis figure will show an isogyre in which the direction of curvature is difficult to determine if $2V$ is very large (> 80°). Olivines from most basic igneous rocks have a compositional range of $Fo_{85}Fa_{15}$ to $Fo_{50}Fa_{50}$, which represents a $2V$ range of 90°–75° (Fig. 2.29).
EXTINCTION	Straight on poor {010} cleavage or prism face.
TWINNING	Rare in most olivines. Sometimes broad deformation lamellae parallel to (100) occur in ultramafic igneous rocks.
OTHER FEATURES	Zoning is occasionally present but is not a diagnostic feature. Mg-rich olivine may contain minor amounts of trivalent ions, which may be exsolved as tiny inclusions of chromite or magnetite.

OCCURRENCE Mg-rich olivine is an essential mineral in most ultrabasic igneous rocks – dunites, peridotites and picrites. Olivine is also common in many basic igneous rocks such as gabbros, dolerites and basalts, and more Fe-rich olivine may occur in some undersaturated igneous rocks such as larvikites, teschenites and alkali basalts. Iron-rich olivine occurs in some ferrogabbros, trachytes and syenites. Olivine can occur in metamorphosed basic igneous rocks and meta-morphosed dolomitic limestones, in which the olivine is nearly pure forsterite.

Olivines may show reaction rims against plagioclase crystals (called kelyphitic margins or corona structures) in metamorphosed basic igneous rocks or in some ultrabasic troctolites, where an olivine or olivine plus serpentine kernel is surrounded by successive rims of pyroxene and garnet or spinel and amphibole. In basic igneous rocks containing Mg-rich olivine, quartz is never present since it would have combined with the olvine in the crystallizing magma to give more orthopyroxene; however, in some later-stage iron-rich igneous rocks (syenogabbros and ferrogabbros), iron-rich olivine coexists with quartz.

Pectolite   Inosilicate

Pectolite $Ca_2NaH[SiO_3]_3$ triclinic
$$1.135:1:0.997$$
$$a = 90°31', \beta = 95°11', \gamma = 102°28'$$

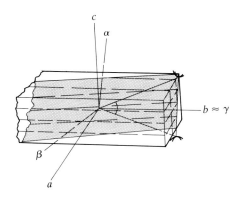

$n_\alpha = 1.595\text{--}1.610$
$n_\beta = 1.604\text{--}1.615$
$n_\gamma = 1.632\text{--}1.645$
$\delta = 0.032\text{--}0.038$
$2V_\gamma = 50°\text{--}63°\ +\text{ve}$
OAP is close to (100) (a prism section is length slow)
$D = 2.75\text{--}2.90 \qquad H = 4\frac{1}{2}$

COLOUR	Colourless.
*HABIT	Pectolite occurs as radiating groups of acicular crystals elongated along the b axis.
*CLEAVAGE	Perfect cleavages on {100} and {001} intersect at about right-angles on an (010) section.
RELIEF	Moderate.
ALTERATION	Pectolite may alter to **stevensite** $Mg_3[Si_4O_{10}](OH)_2$, a sheet silicate, if attacked by solutions containing magnesium bicarbonate.
BIREFRINGENCE	The maximum birefringence, seen in sections perpendicular to the b axis, shows maximum bright interference colours of low third order.
INTERFERENCE FIGURE	Pectolite shows a good Bx_a positive figure on sections cut at right-angles to the prism length.
EXTINCTION	An (010) section shows a small angle fast (α) to cleavage of 5°–11°, with the slow component (β) showing a larger angle of 10°–16°.
OCCURRENCE	Pectolite is deposited by hydrothermal solutions in veins, cavities, and amygdales in basalts and dolerite dykes in association with calcite, zeolites, prehnite and datolite. It is a rare mineral in under-saturated igneous rocks along with feldspathoidal minerals, and may occur in some skarns.

Prehnite   Phyllosilicate

Prehnite $CaAl[AlSi_3O_{10}](OH)_2$ orthorhombic
                                                          $0.843:1:3.378$

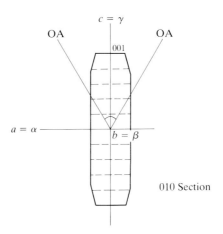

010 Section

$n_\alpha = 1.610–1.637$
$n_\beta = 1.616–1.647$
$n_\gamma = 1.632–1.673$
$\delta = 0.020–0.035$
$2V_\gamma = 60°–70°$ +ve
OAP is parallel to (010) (a prism section is length slow)
$D = 2.90–2.95$ $H = 6–6\frac{1}{2}$

COLOUR   Colourless.
*HABIT   Prehnite occurs as botryoidal masses of radiating crystals. Columnar aggregates of crystals often show a fan-shaped or 'hourglass' structure. Rare single crystals are tabular or prismatic.
*CLEAVAGE   {001} basal cleavage is perfect.
RELIEF   Moderate.
ALTERATION   Prehnite alters to the zeolite scolecite, or chlorite, and may change to epidote or grossular under low-grade metamorphism. Saussuritization of plagioclase may yield prehnite in addition to zoisite.
*BIREFRINGENCE Brilliant interference colours are middle second order, although occasional anomalous colours may be seen.

117

INTERFERENCE FIGURE
: A basal section shows a large positive Bx_a figure.

DISTINGUISHING FEATURES
: The 'hourglass' structure is diagnostic of prehnite, and anomalous interference colours or brilliant second-order colours are peculiar to prehnite.

OCCURRENCE
: Prehnite occurs in amygdales, cavities or veins in basalts, and in other basic extrusive rocks, with zeolites, calcite, chalcedonic silica and pectolite. It is less common in veins and cavities in acid igneous rocks. Prehnite may occur in thermally metamorphosed impure limestones with calcite wollastonite, diopside and grossular. It may also occur as a saussuritization product of plagioclase feldspar along with zoisite.

Pumpellyite   Sorosilicate

Pumpellyite $Ca_2Al_2(Mg,Fe^{2+},Fe^{3+},Al)(SiO_4)(Si_2O_7)(OH)_2(H_2O.OH)$monoclinic
$$1:1:0.842, \beta = 97°36'$$

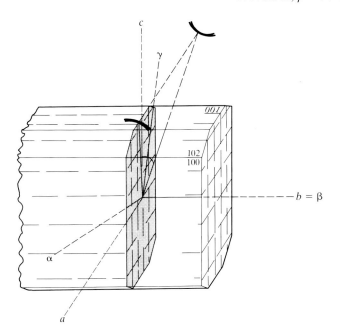

$n_\alpha = 1.674–1.748$
$n_\beta = 1.675–1.754$
$n_\gamma = 1.688–1.764$ } RIS vary with increasing iron content
$\delta = 0.002–0.022$
$2V_\gamma = 10°–85°$ + ve
OAP is parallel to (010)
$D = 3.18–3.23$ $H = 6$

*COLOUR	Green or yellow.
*PLEOCHROISM	a colourless or pale yellow, β pale green or yellow, and γ colourless or yellowish brown.
HABIT	Aggregates of acicular or fibrous crystals, often in rosettes, are common.
CLEAVAGE	{001} and {100} present.
RELIEF	High.
ALTERATION	Uncommon.
BIREFRINGENCE	Low to moderate with variable interference colours from first-order greys to second-order blues.
*INTERFERENCE FIGURE	Highly variable $2V_\gamma$, from 10° to 85° depending on composition, always positive. Near-basal sections give a reasonable Bx_a figure, but a single optic axis figure is needed for iron-rich varieties.
EXTINCTION	Oblique on cleavage, with $\gamma\hat{\ }cl$ and $a\hat{\ }cl$ both $\approx 4°–30°$ seen on a (010) section.
TWINNING	Sector twinning common on {001} and {100}.
DISTINGUISHING FEATURES	Pumpellyite is similar to epidote but has a deeper colour and is biaxial positive (epidote is biaxial negative).
OCCURRENCE	Pumpellyite is common in low-grade regionally metamorphosed schists and is characteristically developed in glaucophane–schist facies rocks. Pumpellyite may form by hydrothermal action in low-temperature veins, amygdales and altered mafic rocks.

Pyroxene group   Inosilicates

Orthorhombic pyroxenes (usually abbreviated to opx) and mono-clinic pyroxenes or clinopyroxenes (abbreviated to cpx) occur. A wide range of compositions is available. The general formula may be expressed as:

$$X_{1-n}Y_{1+n}Z_2O_6$$

where X = Ca,Na, Y = $Mg,Fe^{2+},Ni,Li,Fe^{3+},Cr,Ti$, and Z = Si,Al.

In orthopyroxenes, n is approximately 1 and virtually no trivalent or monovalent ions are present; thus the formula reduces to $Y_2Z_2O_6$ or $(Mg,Fe,Mn)_2(Si,Al)_2O_6$. In clinopyroxenes, n varies from 0 to 1, and the elemental substitutions must be such that the sum of the charges in X, Y and Z balances the six O^{2-} ions.

Some pyroxenes have compositions that can be expressed in terms of $CaSiO_3$–$FeSiO_3$–$MgSiO_3$ end members, **wollastonite**, **clinoferro-silite** and **clinoenstatite**, and the main *monoclinic* pyroxene minerals occurring in this field of composition are shown in Figure 2.30. The principal change occurring in this group is a variation in symmetry between the high Ca-bearing members (> 25% $CaSiO_3$), which are always monoclinic, and the low Ca-bearing members (< 15% $CaSiO_3$). In the latter group the iron-rich members (> 30% $FeSiO_3$) are monoclinic at high temperatures (pigeonites) but invert to an

119

Figure 2.30
The composition
diagram for
pyroxenes.

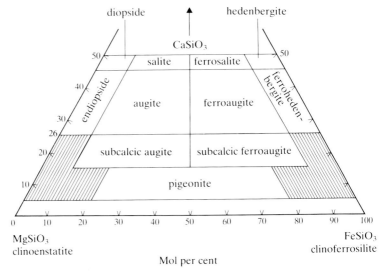

orthorhombic form (opx) at low temperatures. The magnesium-rich members ($< 30\%$ $FeSiO_3$) are orthorhombic at all temperatures. Orthopyroxenes can exist with up to 5 mol% $CaSiO_3$ in the structure.

The remaining pyroxenes of importance cannot be fitted into Figure 2.30 since they include sodium- and aluminium-bearing end members. Two further systems, to include the Na and Al members, are shown in Figure 2.31. In Figure 2.31a the central tie line from Figure 2.30, above $CaMgSiO_3$–$CaFeSiO_3$ (50 mol%) is used as the base of the triangle, and the other apex is $NaFe^{3+}SiO_3$. In Figure 2.31b the tie line $CaMgSiO_3$–$CaFe^{2+}SiO_3$, which represents diopside–hedenbergite or the diopside solid solution series (written as diopside$_{ss}$ or di$_{ss}$), now represents only one corner of the system, and the Na-bearing pyroxene $NaFe^{3+}SiO_3$ another. The third apex of the triangle is the phase $NaAlSiO_3$ or jadeite. The main mineral phases occurring are depicted in the two figures. The shaded area in Figure 2.31a means that a continuous sequence of change from di$_{ss}$ to aegirine ($NaFe^{3+}SiO_3$) does not exist. Similarly, it will be observed from Figure 2.31b that the two main Al-bearing phases jadeite ($NaAlSiO_3$) and omphacite (a cpx containing Na,Al,Fe^{3+} and Mg,Fe^{2+}) are both isolated phases, quite separate from the other end member in the system.

The 'normal' pyroxenes which are represented in Figure 2.30 are essential constituents of the basic calc-alkaline igneous rocks. These pyroxenes may occur in some ultrabasic and intermediate igneous rock types; and some pyroxenes in this system may occur in high-temperature regional and thermal metamorphic rocks.

120

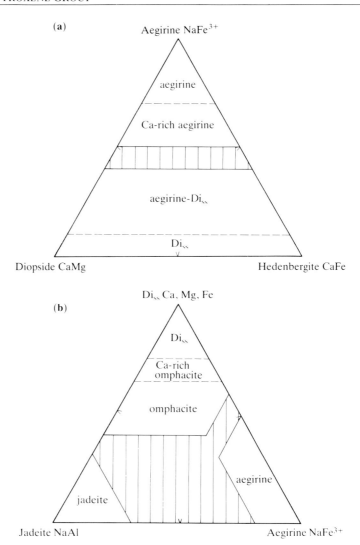

Figure 2.31 The composition diagrams for (a) Na-pyroxenes, and (b) Na- and Al-pyroxenes.

The Na-bearing pyroxenes primarily occur in alkaline igneous rocks of various types, and the Al-bearing pyroxenes occur in rocks (metamorphic or igneous) in which high pressures and temperatures have been operative.

The individual pyroxene minerals (or series) are discussed separately, but extinction angles are given for all the pyroxene minerals in Figure 2.32.

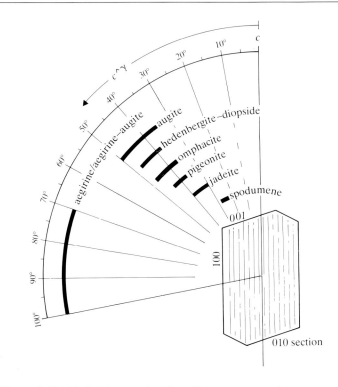

Figure 2.32 Extinction angles for clinopyroxenes: the ranges of $c{\wedge}\gamma$ (maximum extinction angles) for several common pyroxene minerals. Note that $c{\wedge}a$ extinction angles for aegirine and aegirine–augite vary from 0° to 20°.

Exsolution lamellae In many slowly cooled pyroxenes, especially orthopyroxenes and augites, lamellae occur which have a definite crystallographic orientation. These are *not* twin lamellae, but exsolved sheets with a different composition from the host mineral.

Orthopyroxenes may contain lamellae parallel to the (100) plane. These actually are Ca-rich clinopyroxene lamellae; the opx first crystallizes at a high temperature with some calcium in the structure, then cools, and the excess calcium is exsolved as clinopyroxene lamellae parallel to the (100) plane (Fig. 2.33).

Figure 2.33
Exsolution
lamellae parallel
to (100).

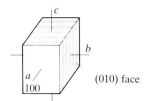

(010) face

122

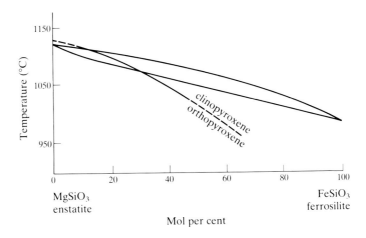

Figure 2.34   The inversion curve and phase diagram for pyroxenes.

As crystallization proceeds, the Mg:Fe ratio in the liquid decreases. When the ratio reaches the value Mg:Fe = 70:30, ortho-pyroxene is replaced by a Ca-poor clinopyroxene (pigeonite). This Ca-poor clinopyroxene may also contain an excess of calcium at high temperature, which will be exsolved as Ca-rich clinopyroxene lamellae as the pigeonite cools. These exsolution lamellae are parallel to the (001) plane of the monoclinic pigeonite. As cooling proceeds, the Ca-depleted pigeonite cools through an "inversion" curve, below which it changes (or inverts) to orthopyroxene (Fig. 2.34). The final result is a crystal of orthopyroxene containing lamellae of Ca-rich clinopyroxene parallel to the monoclinic (001) plane of the original pigeonite (Fig. 2.35).

Ca-rich clinopyroxenes also contain lamellae of exsolved pyroxene parallel to (100) and (001). The lamellae which are parallel to (100) are exsolved orthopyroxene, whereas the lamellae parallel to the (001) plane represent exsolved pigeonite. If cooling proceeds through the inversion curve, the pigeonite lamellae will invert to orthopyroxene, producing a second set of lamellae parallel to (100). Recent research work has shown that the crystallographic orientation of these lamellae, as well as their chemical composition, may be much more complex than was originally thought.

Figure 2.35
Exsolution
lamellae parallel
to (001).

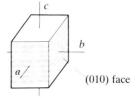

Crystallization trends Two Mg-rich pyroxenes may first crystallize from a basic magma, a Ca-poor orthopyroxene and a Ca-rich augite. As crystallization proceeds, both pyroxenes become more Fe-rich until the Mg:Fe ratio reaches 70:30, at which point pigeonite replaces orthopyroxene as the Ca-poor pyroxene that is crystallizing. Both clinopyroxenes become increasingly Fe-rich as crystallization continues. In many intrusions crystallization ceases at this point, but if fractionation is very marked, only one pyroxene (a Ca-rich ferroaugite) crystallizes when the Mg:Fe ratio drops below 35:65. Finally, in extreme cases, crystallization continues until a Ca-rich ferrohedenbergite appears that contains no Mg. The crystallization sequence, including olivine, can be depicted diagrammatically as follows:

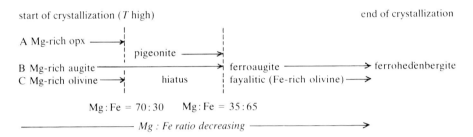

It should be noted that the Mg:Fe ratios given above may change depending upon the amount of cations replacing Mg, Fe (and Ca) in the pyroxene structure. The above diagram can be represented as part of the cpx triangular diagram given in Figure 2.36: A, B and C are shown on the figure.

Orthopyroxenes

Enstatite (En) – orthoferrosilite (Fs)

$MgSiO_3$–$FeSiO_3$
orthorhombic
1.03:1:0.59 (En)

A complete sequence of orthopyroxenes is possible from 100% $MgSiO_3$ (100% enstatite) to 100% $FeSiO_3$ (100% orthoferrosilite). Pure enstatite can be written $En_{100}Fs_0$ and pure orthoferrosilite En_0Fs_{100}. Nowadays, it is conventional to give the composition of an orthopyroxene as the percentage of orthoferrosilite present. Thus Fs_{24} means $En_{76}Fs_{24}$ or, to give its complete formula, $(Mg_{0.76}Fe_{0.24})SiO_3$. The orthopyroxenes can be divided on the basis of composition into several different types, which include **enstatite** (Fs_0–Fs_{12}), **bronzite** (Fs_{12}–Fs_{30}), **hypersthene** (Fs_{30}–Fs_{50}), **ferrohypersthene** (Fs_{50}–Fs_{70}), **eulite** (Fs_{70}–Fs_{88}) and **orthoferrosilite** (Fs_{88}–Fs_{100}). The compositions of the extreme end members, enstatite and

124

Figure 2.36
Crystallization
trends in cpx,
opx and
olivines.

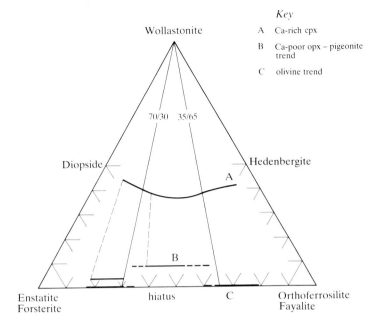

Key

A Ca-rich cpx

B Ca-poor opx – pigeonite
   trend

C olivine trend

Wollastonite

70/30   35/65

Diopside

Hedenbergite

A

B

Enstatite
Forsterite

hiatus

C

Orthoferrosilite
Fayalite

orthoferrosilite, have been selected so that they are optically positive over the complete range.

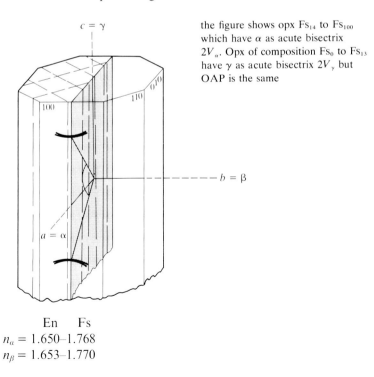

$c = \gamma$

100

110

010

$b = \beta$

$a = \alpha$

the figure shows opx Fs_{14} to Fs_{100}
which have α as acute bisectrix
$2V_\alpha$. Opx of composition Fs_0 to Fs_{13}
have γ as acute bisectrix $2V_\gamma$ but
OAP is the same

En   Fs
$n_\alpha = 1.650–1.768$
$n_\beta = 1.653–1.770$

125

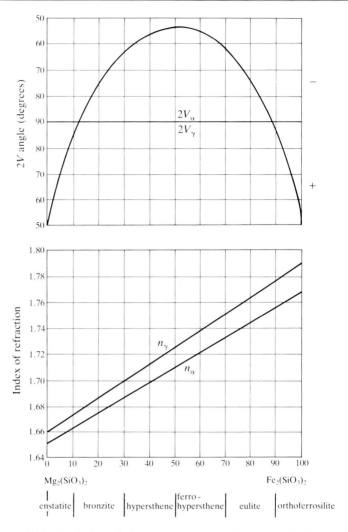

Figure 2.37 Variation of the 2V angle and indices of refraction in the enstatite–orthoferrosilite (orthopyroxene) series.

$n_\gamma = 1.658$–1.788
$\delta = 0.008$–0.020
$2V_\gamma = 60°$–$90°$ + ve (for opx of composition Fs_0–Fs_{13})
$2V_a = 90°$–$50°$ − ve (for opx of composition Fs_{13}–Fs_{87})
High Fe-opx (Fs_{87}–Fs_{100}) are also positive with large 2V, but these are not present in igneous rocks, and only occur in some meteorites. Details of the 2V variation are given in Figure 2.37.
OAP is parallel to (010): it is occasionally shown as being parallel to (100)
$D = 3.21$–3.96 (Fe-rich) $H = 5$–6

COLOUR	Mg-rich opx are colourless; Fe-rich compositions show pale colours, from pale green to pale brown.
PLEOCHROISM	Coloured opx show faint pleochroism with α pink to brown, β yellow to brown, and γ green, due to either Fe^{2+}, Ti or Al in the structure.
HABIT	Early formed opx in igneous rocks appear as short prismatic crystals.
*CLEAVAGE	Two good prismatic {110} cleavages meet at nearly 90° on a basal section. {010} and {100} are poor cleavages or partings.
RELIEF	Moderate to high.
ALTERATION	opx minerals alter to serpentine as follows:

$$3MgSiO_3 + 2H_2O \rightarrow Mg_3(Si_2O_5)(OH)_4 + SiO_2$$

The serpentine mineral is sometimes called **bastite**. Orthopyroxenes may occasionally alter to amphibole, cummingtonite first being formed.

$$8(Mg,Fe)SiO_3 + H_2O \rightarrow (Mg,Fe)_7Si_8O_{22}(OH)_2 + (Fe,Mg)O$$

A rim of amphibole is formed around the opx crystal, and iron ores are released in the reaction, often seen in basic igneous plutonic rocks.

*BIREFRINGENCE	Low with interference colours from low first-order greys (En) to yellow and reds (iron-rich members).
INTERFERENCE FIGURE	Large biaxial figures are seen on (100) sections.
*EXTINCTION ANGLE	All opx have straight extinction on prism edge or main prismatic cleavages.
TWINNING	Absent from opx.
*OTHERS	One set of exsolution lamellae is usually present, parallel to prismatic face (100). Another set may be present at a high angle to this. The explanation has been given under the heading "Exsolution lamellae".
DISTINGUISH-ING FEATURES	Orthopyroxenes are distinguished from other clinopyroxenes by their parallel extinction. Opx is length slow, whereas andalusite is length-fast; sillimanite, although length-slow, has a very small $2V$ and higher interference colours.
OCCURRENCE	Orthopyroxenes occur in basic igneous rocks of all types. Mg-rich opx occurs in ultrabasic igneous rocks such as pyroxenites, harzburgites, lherzolites and picrites, in association with Mg olivine, Mg augite and Mg spinel. Orthopyroxenes occur in some regional metamorphic rocks, particularly charnockites and granulites, and may occur at high grades during the thermal metamorphism of argillaceous rocks in hornfelses of the innermost zones of thermal aureoles.

Clinopyroxenes

Diopside (Di) – hedenbergite (Hed)

$CaMgSi_2O_6$–$CaFeSi_2O_6$ monoclinic

Di $1.091:1:0.589, \beta = 105°51'$

Hed $1.091:1:0.584, \beta = 105°25'$

A series of minerals exist between diopside and hedenbergite, including **diopside** (Di_{100}–Di_{90}), **salite** (Di_{90}–Di_{50}), **ferrosalite** (Di_{50}–Di_{10}) and **hedenbergite** (Di_{10}–Di_0).

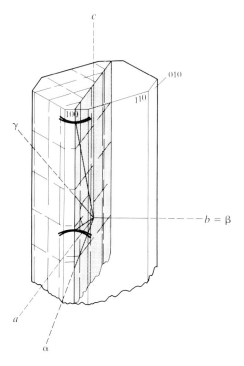

$n_\alpha = 1.664$–1.726

$n_\beta = 1.672$–1.730

$n_\gamma = 1.694$–1.751

$\delta = 0.030$–0.025

$2V_\gamma = 58°$–$63°$ + ve

OAP is parallel to (010)

$D = 3.22$–3.56 $H = 5\frac{1}{2}$–$6\frac{1}{2}$

COLOUR   Diopside is colourless and hedenbergite is brownish green.

PLEOCHROISM   Hedenbergites are pleochroic in pale greens and browns, but pleochroism is weak and is not a diagnostic feature.

HABIT   Occurs as short subhedral prisms.

*CLEAVAGE   {110} good prismatic cleavages meeting on a basal section at 87°.

128

Several partings {100}, {010} and {001} are present. A diopside with the {100} parting well developed is called **diallage**.

RELIEF Moderate to high.

ALTERATION Similar to that of the orthopyroxenes. Diopsides will alter commonly to fibrous tremolite–actinolite, and rarely to chlorite.

*BIREFRINGENCE Moderate, middle second order greens and yellows common.

INTERFERENCE FIGURE Moderate $2V$ shown on a (100) section.

*EXTINCTION ANGLE Large extinction angle seen in an (010) section. The angle γ (slow ray)^cleavage is variable to over 45°. Extinction angles for all pyroxenes are shown in Figure 2.32.

TWINNING Simple and multiple twins common on {100} and {001}.

OCCURRENCE Diopside occurs in a wide variety of metamorphic rocks, particularly metamorphosed dolomitic limestones and calcium-rich sediments. Diopside forms from the breakdown of the amphibole tremolite as the temperature increases. Diopside is usually accompanied by forsteritic olivine and calcite. Hedenbergite occurs in metamorphosed iron-rich sediments, being found in eulysites and skarns. Diopside may also occur in some basic extrusive igneous rocks, and hedenbergite may appear in some acid igneous rock types such as fayalite granite, fayalite ferrogabbro and some granophyres.

Pigeonite $(Ca,Mg,Fe)(Mg,Fe)Si_2O_6$ monoclinic

$\beta = 108°$

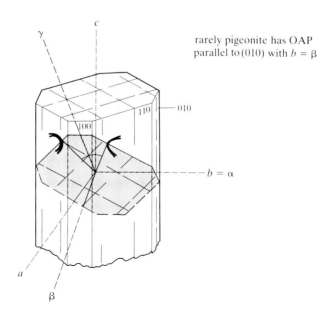

rarely pigeonite has OAP parallel to (010) with $b = \beta$

$n_\alpha = 1.682–1.722$
$n_\beta = 1.684–1.722$
$n_\gamma = 1.705–1.751$
$\delta = 0.023–0.029$
$2V_\gamma = 0°–32°$ + ve
OAP is perpendicular to (010)
$D = 3.30–3.46$ $H = 6$

*OPTICAL
PROPERTIES
Very similar to diopside and augite, except for $2V$ which is small ($< 30°$). A section cut parallel to face (101) will show a good Bx_a figure. The section is difficult to recognize, but the interference colours will be very low (first-order greys) and two cleavages will be present, meeting at an angle of less than 90°.

OCCURRENCE
Pigeonite occurs only in rapidly chilled rocks. In most igneous rocks which have undergone slow cooling, pigeonite is inverted to ortho-pyroxene.

Augite (ferroaugite)

$Ca(Mg,Fe,Mn,Fe^{3+}, Al,Ti)_2(Si,Al)_2O_6$ monoclinic
$1.092:1:0.584, \beta = 105°50'$

Fassaite is a variety of augite in which appreciable substitution of Al and Fe_{3+} for Mg and Fe^{2+} is coupled with substitution of Al for Si.

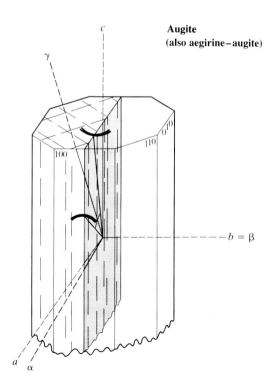

Augite
(also aegirine–augite)

$n_\alpha = 1.662–1.735$
$n_\beta = 1.670–1.741$
$n_\gamma = 1.688–1.761$
$\delta = 0.018–0.033$
$2V_\gamma = 25°–60° + ve$

Refractive indices change depending upon the Mg:Fe ratio, and also on the amount and lattice position of the minor constituents Al, Ti and Fe^{3+}. For example, Al (Ti and Fe) in tetrahedral co-ordination (occupying Si sites) will increase $2V$ and lower RIs, whereas in the octahedral sites (occupying Mg and Fe positions) they will decrease $2V$ and increase RIs. These factors may affect compositions determined from optical properties by $\sim 5\%$.

OAP is parallel to (010)

$D = 2.96–3.52$ $H = 5–6$

COLOUR
Augite is colourless to pale brown. Titanaugite (Ti-augite) is pale purple.

PLEOCHROISM
In most varieties pleochroism is very weak, but titanaugite is weakly pleochroic with α pale green, β pale brownish and γ pale greenish purple.

HABIT
Variable, from subhedral prismatic crystals in basic plutonic rocks to euhedral crystals in basic extrusive rocks.

*CLEAVAGE
Similar to diopside, with {110} good and poor {100} {010} partings.

RELIEF
Moderate to high.

ALTERATION
Similar to diopside.

BIREFRINGENCE
Moderate, with maximum interference colours being low second-order (blues and greens).

*INTERFERENCE FIGURE
A good Bx_a figure is seen on the plane including the a and c axes, about the position of the face (101).

*EXTINCTION ANGLE
Similar to diopside; $\gamma{\char`\^}cl$ large (up to $45° +$).

TWINNING
Similar to diopside.

*OTHERS
Hourglass zoning is occasionally seen on certain prismatic sections, especially in titanaugite.

DISTINGUISHING FEATURES
Augite is virtually indistinguishable from diopside. It may show a slightly smaller $2V$. Augite is found in mafic and ultramafic plutonic igneous rocks, whereas diopside occurs mostly in metamorphic rocks and basic volcanics.

OCCURRENCE
Augites occur mainly in igneous rocks and are essential mineral constituents of gabbros, dolerites and basalts. In plutonic gabbros augites frequently occur with orthopyroxenes (as already described under "Crystallization trends"). Augitic cpx have been recognized in some very high-grade metamorphic rocks (granulites).

Jadeite $NaAlSi_2O_6$

monoclinic
$1.103 : 1 : 0.613, \beta = 107°20'$

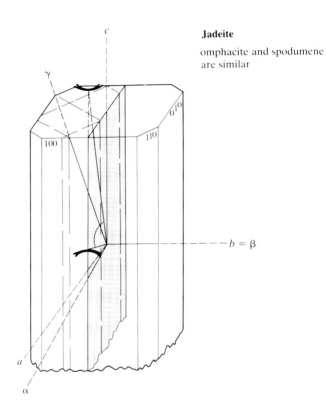

Jadeite

omphacite and spodumene
are similar

$n_\alpha = 1.640–1.658$
$n_\beta = 1.645–1.663$
$n_\gamma = 1.652–1.673$
$\delta = 0.012–0.015$
$2V_\gamma = 70°–75°$ + ve
OAP is parallel to (010)
$D = 3.24–3.43$ $H = 6$

COLOUR	Colourless.
HABIT	Usually occurs as granular aggregates of crystals, usually medium- to coarse-grained.
CLEAVAGE	Similar to diopside.
RELIEF	Moderate.
ALTERATION	Jadeite can alter to amphiboles, or dissociate to a mixture of nepheline and albite.
BIREFRINGENCE	Moderate, with second-order colours.
EXTINCTION ANGLE	Similar to diopside, but smaller maximum angle with γ (slow)ˆcleavage = 33°–40° (max).

132

*OCCURRENCE   Jadeite is a rare pyroxene which can sometimes occur with albite in regional metamorphic rocks, especially glaucophane schists, which have formed under low heatflow, high P conditions. It is found in metamorphic rocks in association with lawsonite and glaucophane.

Omphacite   This is a cpx which has properties similar to jadeite and augite, and is found exclusively in eclogites. In this rock, omphacite occurs with an Mg,Fe-garnet and the assemblage is considered to have formed at high T and P, and to be stable at temperatures below 700°C with pressures of > 10 kb. The higher the pressure of formation, the richer the omphacite is in Na and Al.

Aegirine **(acmite)**	$NaFe^{3+}Si_2O_6$	$1.099:1:0.601, \beta = 106°49'$ axial ratio variable
Aegirine– **augite**	$(Na,Ca)(Fe,Fe^{3+},Mg)Si_2O_6$	

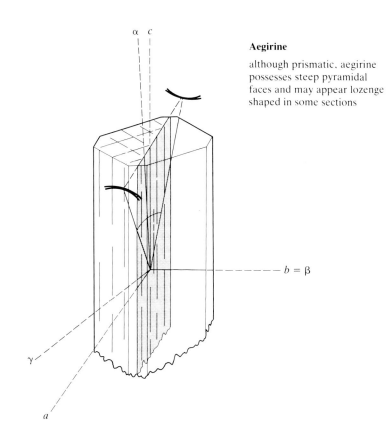

Aegirine

although prismatic, aegirine possesses steep pyramidal faces and may appear lozenge shaped in some sections

Aegirine	Aegirine–augite
$n_a = 1.750–1.776$	$n_a = 1.700–1.750$
$n_\beta = 1.780–1.820$	$n_\beta = 1.710–1.780$
$n_\gamma = 1.800–1.836$	$n_\gamma = 1.730–1.800$
$\delta = 0.040–0.06$	$\delta = 0.03–0.05$
$2V_a = 58°–90°$ − ve	$2V_\gamma = 90°–70°$ + ve
OAP is parallel to (010)	OAP is parallel to (010)
$D = 3.55–3.60$ $H = 6$	$D = 3.40–3.55$ $H = 6$

*COLOUR — Aegirine is strongly coloured in shades of green. Aegirine–augite is also coloured green in thin section.

*PLEOCHROISM — Aegirine is strongly pleochroic, with a emerald green, β deep green and γ brownish green. Aegirine–augite is also pleochroic with a similar scheme to that of aegirine, but usually has a more yellowish colour throughout.

HABIT — Elongate prismatic crystals are most common.

CLEAVAGE — Similar to other pyroxenes.

RELIEF — High to very high.

*BIREFRINGENCE — High but third-order interference colours are usually masked by a strong greenish colour, which persists under crossed polars.

INTERFERENCE FIGURE — $2V$ data are given in Fig. 2.38.

*EXTINCTION ANGLE — Both aegirine and aegirine–augite have small extinction angles in an (010) prismatic section. The extinction angles a (fast)ˆc axis (or aˆprismatic cleavage) vary from 0° to ∼ 20°.

*OCCURRENCE — Aegirine and aegirine–augite occur as late crystallization products of alkali magmas, appearing in syenites and nepheline–syenites with alkali amphiboles (see earlier section in this chapter). They may occur in alkali granites, often with riebeckite, and may occur in some Na-rich schists with glaucophane and riebeckite.

Spodumene $LiAlSi_2O_6$ monoclinic
                                   $1.144:1:0.632, \beta = 110°20'$

$n_a = 1.648–1.663$
$n_\beta = 1.655–1.669$
$n_\gamma = 1.662–1.679$
$\delta = 0.014–0.027$
$2V_\gamma = 54°–69°$ + ve
OAP is parallel to (010)
$D = 3.03–3.22$ $H = 6\frac{1}{2}–7$

The properties of spodumene are similar to those of diopside (colourless in thin section and so on), but spodumene is a rare mineral, occurring in lithium-rich acid igneous rocks such as granite pegmatites, where it is associated with quartz, albite, lepidolite (lithium-rich mica), beryl and tourmaline.

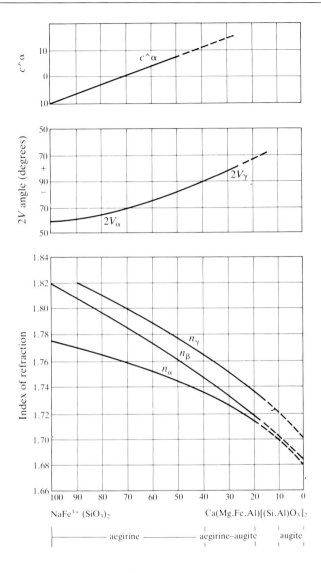

Figure 2.38 Variation of the $c^\wedge a$, $2V$ angle and indices of refraction in the aegirine–augite series (data from Deer, Howie & Zussman 1962).

Pseudo-pyroxenes

Wollastonite $CaSiO_3$ triclinic

$$1.082 : 1 : 0.965$$
$$a = 90°02', \beta = 95°22', \gamma = 103°26'$$

Although used as an end member of the pyroxene group of minerals

135

(Fig. 2.30), wollastonite does not possesss a pyroxene structure, but is chemically similar and is described with them.

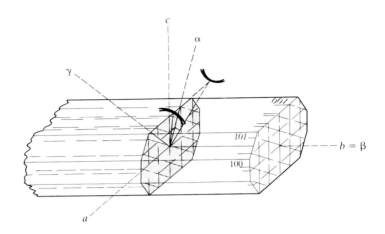

$n_\alpha = 1.616$–1.640
$n_\beta = 1.628$–1.650
$n_\gamma = 1.631$–1.650
$\delta\ = 0.013$–0.014
$2V_\alpha = 38°$–$60°\ -$ve
OAP is approximately parallel to (010)
$D = 2.87$–3.09 $\quad H = 4\frac{1}{2}$–5

COLOUR	Colourless.
HABIT	Usually columnar or fibrous, with rectangular cross sections.
*CLEAVAGE	{100} perfect and {001} and {10$\bar{2}$} good. A typical section shows two or three cleavages.
RELIEF	Moderate.
BIREFRINGENCE	Low, with maximum first-order yellow.
*INTERFERENCE FIGURE	Biaxial negative, with a moderate $2V$, seen almost on the basal face (i.e. at right-angles to the length of the crystals).
EXTINCTION ANGLE	Since the cleavages are different from those of pyroxene, the extinction angle is not so relevant a feature, but the crystals are almost length-fast, and $a\hat{\ }c$ axis (crystal length) is 30°–40°.
DISTINGUISHING FEATURES	Identification of wollastonite is difficult. However, it should be noted that although wollastonite is virtually identical to diopside, it is optically negative, whereas diopside is optically positive.
OCCURRENCE	Wollastonite is a mineral formed in metamorphosed impure (calcareous) limestones, usually as a result of the reaction

$$CaCO_3 + SiO_2 \text{ (impurity)} \rightarrow CaSiO_3 + CO_2$$

at fairly high temperatures (about 1000°C), which may be reduced if volatiles are present. Wollastonite has been recorded from some alkaline igneous rocks.

Parawollastonite (the monoclinic form of wollastonite) has a similar paragenesis.

Sapphirine   Nesosilicate

Sapphirine $(Mg,Fe)_2Al_4O_6[SiO_4]$?monoclinic

$0.69:1:0.70, \beta = 125°\,20'$

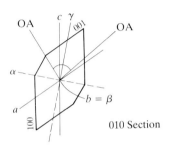

010 Section

$n_\alpha = 1.701–1.729$
$n_\beta = 1.703–1.732$
$n_\gamma = 1.705–1.734$
$\delta = 0.004–0.007$
$2V_\gamma = 66°–90°$ + ve; $2V_\alpha = 90°–50°$ − ve (sapphirine may be + ve or − ve)
OAP is perpendicular to the *b* axis
$D = 3.40–3.58$ $H = 7\frac{1}{2}$

COLOUR   Colourless to pale blue.
*PLEOCHROISM   Distinct with *a* colourless, pale yellow or greenish blue, *β* pale blue, bluish-green or pale violet blue, and *γ* deep blue or green.
HABIT   Euhedral crystals are elongated along the *c* axis; crystal aggregates are common.
CLEAVAGE   Poor cleavages on {100}, {010} and {001}.
RELIEF   High.
ALTERATION   Sapphirine alters to a mixture of corundum, biotite and perhaps talc.
*BIREFRINGENCE First-order greys common, but anomalous indigo greys may occur.
INTERFERENCE   Sections normal to {010} show Bx_a figures. A single optic axis should
FIGURE   be obtained for the sign.

EXTINCTION	Oblique extinction is seen on all figures.
*OCCURRENCE	Sapphirine is a high-temperature mineral of both regional and thermal metamorphism. It is characteristic of granulite amphibolite facies and Mg-rich, silica-poor rocks. It is found in schists, gneisses, granulites and hornfelses, and is associated with corundum, cordierite, spinel, diaspore, sillimanite, anthophyllite, hypersthene and anorthite. In emery deposits it occurs with corundum, magnetite and spinel. It is not compatible with quartz, olivine, enstatite and periclase.

Scapolite   Tektosilicate

Scapolite $(Ca,Na)_4[(Al,Si)_3Al_3Si_6O_{24}](Cl,CO_3)$ tetragonal c/a 0.620

$n_o = 1.540–1.600$ ⎫ Indices increase with increasing substitution of
$n_e = 1.535–1.565$ ⎭ Ca for Na
$\delta = 0.004–0.037$
Uniaxial $-$ ve (length fast)
$D = 2.50–2.78$ $H = 5–6$

COLOUR	Colourless.
*HABIT	Large spongy prismatic crystals are common in metamorphosed carbonate rocks. Granular and fibrous-looking aggregates are also common, especially in garnet-bearing rocks.
CLEAVAGE	Good {100} and {110} prismatic cleavages.
RELIEF	Low.
ALTERATION	Scapolite may alter to a fine aggregate containing combinations of various minerals including chlorite, sericite, epidote, calcite, plagioclase and clays.
BIREFRINGENCE	Low (Na-varieties) to moderate (Ca-varieties), with interference colours varying accordingly.
INTERFERENCE FIGURE	Aggregates of crystals are usually too small for sign determination.
ZONING	Compositional zoning is common.
DISTINGUISHING FEATURES	Na-rich scapolite is similar in RI and birefringence to quartz, K-feldspar, plagioclase and cordierite, but it is uniaxial negative and untwinned, whereas quartz is uniaxial positive. K-feldspars and cordierite are biaxal, and plagioclase is biaxial and invariably twinned. Nepheline is also uniaxial negative but its RIS are lower (often $<$ CB) and it has a different occurrence.
OCCURRENCE	Scapolite may occur in some pegmatites, replacing plagioclase or quartz, but its main occurrence is in metamorphic or metasomatic rocks.
	Scapolite may form as a primary mineral in calcareous rocks subjected to regional metamorphism at amphibolite facies. It is associated with calcite, sphene, diopside, plagioclase, epidote and

garnet. In contact metamorphism, scapolite forms in carbonate rocks by introduction of sodium and chlorine from the igneous intrusion. Grossular garnet, wollastonite and fluorite are associated minerals.

Serpentine   Phyllosilicate

Serpentine $Mg_3Si_2O_5(OH)_4$ monoclinic

$0.57:1:1.31, \beta = 93°$

Serpentine includes a variety of minerals, one fibrous (**chrysotile**) and two tabular (**lizardite** and **antigorite**)

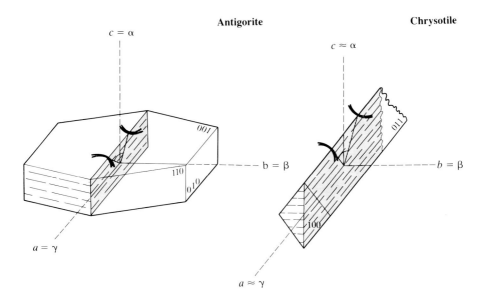

	Chrysotile	Lizardite	Antigorite
n_a	1.53–1.55	1.54–1.55	1.56–1.57
n_β	—	—	1.57
n_γ	1.55–1.56	1.55–1.56	1.56–1.57
δ	0.013–0.017	0.006–0.008	0.004–0.007
$2V_a$	variable − ve	?	37°–61° − ve
OAP	parallel to (010)	?	parallel to (010)
D	2.55	2.55	2.6
H	$2\frac{1}{2}$	$2\frac{1}{2}$	2–$3\frac{1}{2}$

COLOUR   Colourless to pale green.

HABIT   Chrysotile is fibrous elongated parallel to the a crystallographic axis, and lizardite and antigorite are both flat, tabular crystals.

CLEAVAGE   Chrysotile has a fibrous cleavage, and lizardite a basal cleavage.

139

RELIEF	Low.
*BIREFRINGENCE	Low or very low, often with anomalous pale yellow colours shown (cf. chlorite).
INTERFERENCE FIGURE	Antigorite shows a medium sized negative $2V$ on a basal section.
EXTINCTION	Straight on fibres, cleavage or crystal edge. Chrysotile is length-slow.
DISTINGUISHING FEATURES	Serpentine minerals have a lower birefringence and lower refractive indices than chlorite and fibrous amphiboles. Most chlorites exhibit either stronger birefringence or anomalous interference colours. Brucite can show anomalous colours similar to chlorite, but brucite is uniaxial.
*OCCURRENCE	Serpentine minerals are formed during the alteration of ultrabasic igneous rocks – dunites, pyroxenites and peridotites – at temperatures below 400°C. Chrysotile probably forms first and antigorite is then derived from it.

Chrysotile is the major variety of commercial asbestos and it occurs as economic deposits in Canada, South Africa and Russia.

Silica group   Tektosilicates

The various forms of silica (SiO_2 is the formula for all silica minerals) can be represented on a simple $P - T$ diagram (Fig. 2.39). This shows that the lowest temperature form of quartz, called a-**quartz** (or low quartz), inverts to β-**quartz** (or high quartz) at 573°C at atmospheric pressure; the temperature of this inversion increases with increasing pressure (~ 670°C at 3 kb). At 867°C, β-quartz inverts to **tridymite**; the temperature of this inversion also increases considerably with increasing pressure (to ~ 1450°C at 3 kb). Tridymite inverts to **cristobalite** at 1470°C at atmospheric pressure, and the temperature of this inversion does not change with increasing pressure. Finally, at 1713°C cristobalite melts and the liquidus boundary is reached. This diagram can, of course, be interpreted in the other direction, with liquid SiO_2 crystallizing: one can then determine which polymorph (minerals with the same chemistry but a different structure) will be encountered at which temperature as crystallization proceeds.

At very high pressures, two other structural phases are known, namely **coesite** and **stishovite** (see Fig. 2.39), both of which are recorded from meteorite impact sites; but these minerals rarely occur in normal terrestrial rocks. However, other minerals possessing the same type of atomic lattice as stishovite may exist in the Earth's upper mantle. Coesite has been recorded from the Alps as inclusions in pyrope garnet in gneisses, which must have formed at a pressure of 30 kb (3 GPa) at depths in excess of 85 km. Quartz is an essential constituent of acid igneous rocks and arenaceous sedi-

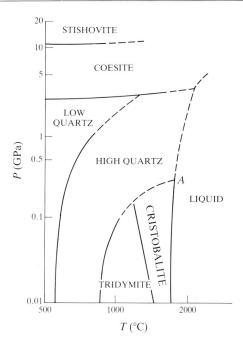

Figure 2.39 The P–T stability fields of the various SiO_2 polymorphs. Note that 1 kb = 0.1 GPa

mentary rocks, and is a common constituent of metamorphic rocks. The three main quartz minerals are described.

Quartz	SiO_2	trigonal, c/a 1.100

$n_o = 1.544$
$n_e = 1.553$
$\delta = 0.009$
Uniaxial + ve (length slow)
$D = 2.65 \qquad H = 7$

*COLOUR	Colourless.
HABIT	Euhedral quartz crystals are prisms with hexagonal cross sections, and may appear as phenocrysts in acid extrusive rocks, but quartz usually occurs as shapeless interstitial grains in igneous and meta-morphic rocks, or as rounded grains in sedimentary clastic rocks.
*CLEAVAGE	None.
RELIEF	Low, just greater than 1.54.
*ALTERATION	None.
*BIREFRINGENCE	Low, maximum interference colours are first-order white or pale yellow.
EXTINCTION	Straight on prism edge.

141

TWINNING Many types of twins occur, in particular Brazil (twin plane $11\bar{2}0$) and Dauphiné (twin axis is c axis), but twinning cannot be detected under the microscope because the optic orientation in both twin parts is identical in both types of twin.

OTHERS In some porphyritic acid extrusive and hypabyssal igneous rocks where quartz occurs as phenocrysts, the crystals may show corroded margins because of a reaction between the quartz and the magmatic liquid.

OCCURRENCE Quartz is an essential mineral in acid igneous plutonic rocks such as granites and granodiorites, but may be present in diorites and some gabbros. In these, quartz occurs as shapeless grains. In rapidly cooled extrusive and hypabyssal rocks, rhyolites, dacites, pitchstones and quartz porphyries for example, quartz may occur as euhedral phenocrysts. It is also found as large late-formed crystals in pegmatites, and is a common constituent of hydrothermal veins accompanying various ore minerals. Quartz is one of the most common detrital minerals owing to its lack of cleavage, its hardness and its stability. Because of this, quartz is a common and often essential mineral in coarse terrigeneous rocks such as conglomerates and sandstones, and also occurs in siltstones and mudstones, where its fine grain size is such that detection may not be possible with a microscope. In many sedimentary rocks, including some limestones, authigenic quartz will form. The secondary quartz actually grows around pre-existing grains or forms well developed crystals, and is produced during diagenesis after deposition of the sediments. Quartz occurs in many metamorphic rocks, usually remaining in the rocks until high grades of metamorphism are reached. At these highest grades a reaction

$$\text{muscovite } + \text{ quartz } \rightarrow \text{K-feldspar } + \text{ sillimanite}$$

may occur, and the grain size of these high-grade gneisses will be about the same as that of granite (greater than 3 mm).

Tridymite SiO_2 orthorhombic

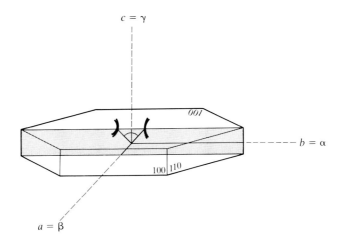

$n_\alpha = 1.469–1.479$
$n_\beta = 1.470–1.480$
$n_\gamma = 1.473–1.483$
$\delta = 0.004$
$2V_\gamma = 40°–90° \ +ve$
OAP is parallel to (100)
$D = 2.26$ $H = 7$

COLOUR   Colourless.
*HABIT   Six-sided tabular crystals common, but usually very tiny.
CLEAVAGE   A poor prismatic cleavage occurs.
RELIEF   Moderate, RI considerably less than 1.54.
ALTERATION   None.
BIREFRINGENCE Low: very low first-order colours.
INTERFERENCE   Very difficult to obtain because of very small crystal size.
FIGURE
*TWINNING   Common on {110}, seen as wedge shaped (sector type) twinning on basal plane.
OCCURRENCE   Rare in rocks, although tridymite may be found in quickly cooled igneous rocks such as rhyolites, pitchstones and dacites. It may be found in association with sanidine (see under "Feldspar group") and sometimes augite or fayalitic olivine. Tridymite has been recorded from high-temperature thermally metamorphosed impure limestones.

Cristobalite SiO_2 tetragonal, c/a 1.395

143

$n_o = 1.478$
$n_e = 1.484$
$\delta = 0.003$
Uniaxial $-$ ve
$D = 2.38$ $\quad$ $H = 6–7$

COLOUR	Colourless.
*HABIT	Minute square-sectioned crystals are common, but cristobalite often occurs as skeletal fibrous crystals in cavities.
CLEAVAGE	None.
RELIEF	Moderate, considerably less than 1.54.
ALTERATION	None.
BIREFRINGENCE	Very weak.
INTERFERENCE FIGURE	Basal sections give a uniaxial negative figure, but the minute crystal size makes this difficult to obtain.
TWINNING	Twins on {111}, but not seen in thin section.
OCCURRENCE	Usually found in cavities in volcanic rocks, and has been discovered in some thermally metamorphosed sandstones. Since both tridymite and cristobalite occur as metastable forms outside their stability field, no conclusions can be drawn about the P–T conditions of formation of rocks containing these minerals.

Cryptocrystalline silica

$SiO_2.nH_2O$

Most varieties are mixtures of cryptocrystalline silica and hydrous silica, from SiO_2 (chalcedonic silica) to opal (hydrous silica).

Chalcedony $\qquad\qquad$ **Opal**
$n_o = 1.526–1.544$ $\qquad$ $n = 1.435–1.460$
$n_e = 1.531–1.533$
$\delta = 0.005–0.009$
Uniaxial $+$ ve
$D = 2.50–2.67$ $\qquad\quad$ $D = 1.99–2.25$
$H = 6\frac{1}{2}–7$ $\qquad\qquad$ $H = 5\frac{1}{2}–6\frac{1}{2}$

In chalcedony, RIs and birefringence decrease with increasing water content. All types are colourless, with some banding or zoning present.

Chalcedony includes a number of subvarieties based mainly on colour. **Agate** is a variegated chalcedony composed of different coloured concentric bands, with sharp or diffuse boundaries. **Moss agate** is a chalcedony containing small dendrites (tree-like growths), which consist of iron oxide or an iron-rich chlorite. **Onyx** is a flat banded variety, having white and grey or brown or black bands. **Flint** is usually black or shades of grey in colour, and is found as nodules, occurring as bands in the Upper Chalk of England. **Chert** is

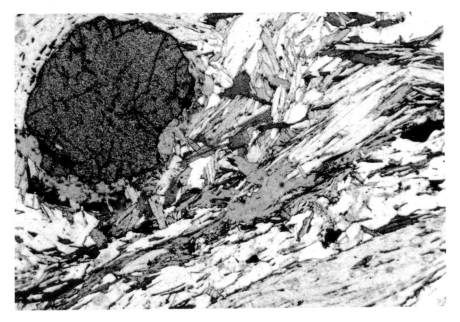

(a)

(b)

Plate 1   Garnet schist. (a) Garnet schist in thin section (PPL). Note the abundant muscovite laths (transparent, moderate relief), generally oriented NE–SW, associated with small rounded grains of quartz (transparent, low relief), chlorite laths (green) and biotite laths (variable brown due to pleochroism). In the top left, the large round crystal with high relief is garnet. There are two small opaque grains towards the right of the image. **(b)** Garnet schist (XPOLS). The muscovite displays second-order interference colours (mainly yellows), while the biotite also displays second-order colours, but the chlorite has a dark brownish-grey interference colour. The quartz has variable first-order white to black interference colours. The garnet is black: it is isotropic and it remains black on rotation of the stage. Some of the other transparent minerals also appear black, but these are in extinction orientations and will reveal their interference colours when the stage is rotated.

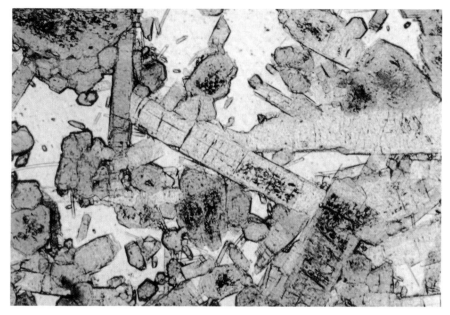

(a)

(b)

Plate 2 Tourmaline crystals. (a) Tourmaline crystals in thin section (PPL). The variation in the bluish-green colour is due to pleochroism. Compositional zoning results in the centres of some crystals being coloured slightly brownish. Note that elongated crystals oriented north–south and rounded basal sections (cut across the acicular mineral) give the deepest colour. The elongated sections vary most in colour (pleochroism) on rotation of the stage while the basal sections maintain their colour. Thus the *mineral* is said to be distinctly pleochroic, even though individual grains exhibit this to varying degrees, depending on their crystallographic orientation. **(b)** Tourmaline crystals (XPOLS). The interference colours range up to second order yellow. Note the variety of colours, mainly due to the different crystallographic orientation of each grain, but also due to variation in thickness of the grains, which tend to thin towards the edges. Grains oriented north–south and east–west are in the extinction orientation and appear black (invisible against the black isotropic background matrix). Basal sections are also black, and remain so on rotation of the stage: they are isotropic. Near-isotropic sections are grey.

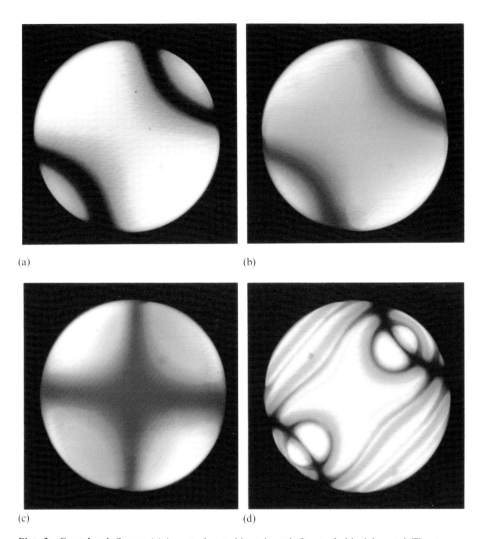

(a)

(b)

(c)

(d)

Plate 3 Crystal optic figures. (a) A centred acute bisectrix optic figure of a biaxial crystal. The stage is rotated 45° from the extinction orientation of the mineral, where a black cross is produced. The curvature and separation of the isogyres which is at a maximum at the 45° orientation, corresponds to a $2V$ of about 40° **(b)** A centred acute bisectrix optic figure of a biaxial crystal, with the sensitive tint inserted in order to determine the optic sign. Since the tint is orientated with its slow vibration direction running NE–SW, and the colour on the concave side of the isogyres is yellow, then the mineral is optically negative. If the mineral had been optically positive then the colour on the concave side of the isogyres would have been blue and the central area yellow. **(c)** A centred optic axis figure of a uniaxial mineral. The sensitive tint is inserted with the slow vibration direction oriented NE–SW. The mineral is optically positive since the blue colour appears in the top right and bottom left quadrants. The violet colour of the cross represents the sensitive tint interference colour at the top of the first order of Newton's Scale. **(d)** A centred acute bisectrix optic figure of a mineral which has a large birefringence, which leads to the greater display of Newton's Scale of Colours.

Plate 4 Transmitted- and reflected-light images of mineral sections. (a) A transmitted-light (PPL) image of a polished thin section of lead ore. Quartz is transparent, while iron-poor sphalerite shows only minor absorption of light, to give a brownish yellow colour. Most of the ore is opaque (black). **(b)** A reflected-light (PPL) image of the lead ore. Quartz is very dark grey. Sphalerite is light grey, while galena is white and displays its typical black triangular cleavage pits. Note the yellow chalcopyrite. **(c)** A reflected-light (PPL) image of a polished section of coarse-grained marcasite. **(d)** A reflected-light (XPOLS) image of marcasite. Crossed-polar observation reveals the distinct aniso-tropy, as well as the multiple twinning, of marcasite. **(e)** A reflected-light (PPL) image of a polished section of mercury ore. Quartz is very dark grey. Cinnabar is light grey and varies in brightness; it is bireflecting, and this variation in brightness is more obvious on rotation of the stage. **(f)** A reflected-light (XPOLS) image of the mercury ore. The main feature is the bright red colour due to internal reflections from the cinnabar. Cinnabar transmits red light, while the quartz is mainly transparent but also "glows" red due to the surrounding cinnabar. **(g)** A reflected-light (XPOLS) image of a polished section of covellite, with the polars exactly crossed. Note the grey and pale brown polarisation colours. **(h)** A reflected-light (XPOLS) image of covellite, but with the analyzer rotated a few degrees. The dispersion of the angle of rotation of the reflected light results in reddish and bluish grey polarization colours.

a grey to black opaque variety which resembles flint, occurring as nodules and beds in the Carboniferous limestone of North Wales, and often being associated with black shales and spilites. **Jasper** is an impure, opaque variety, red, brown, or yellow in colour, and opaque, even on thin edges. **Opal** is the gem variety, exhibiting opalescence and a brilliant play of colours. **Diatomite** or *kieselguhr* is a deposit from the tests or skeletons of siliceous organisms, such as algæ and diatoms, forming beds in lakes, and thick deposits where siliceous volcanic emanations have supplied abundant material for diatom growth, as in the Miocene beds of California.

Sphene   Nesosilicate

Sphene, Titanite

$CaTiSiO_4(O,OH,F)$ monoclinic

$0.755:1:0.854, \beta = 119°43'$

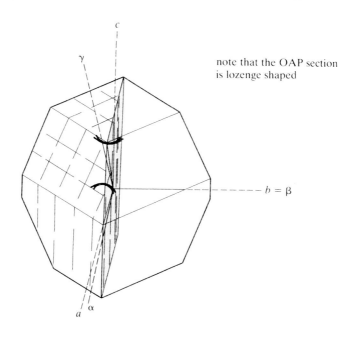

note that the OAP section is lozenge shaped

$n_\alpha = 1.843–1.950$
$n_\beta = 1.879–2.034$
$n_\gamma = 1.943–2.110$
$\delta = 0.100–0.192$
$2V_\gamma = 17°–40°$ (usually 23°–34°) + ve
OAP is parallel to (010)
$D = 3.45 \quad H = 5$

COLOUR	Colourless, pale brown or dark brown.
PLEOCHROISM	Well displayed in sphenes from alkali igneous plutonic rocks, with α yellow or colourless, β pinkish or yellowish brown, γ pink, yellow, orange or brown.
*HABIT	Anhedral to euhedral, often occurring as small lozenge- or diamond-shaped crystals.
*CLEAVAGE	{110} good.
*RELIEF	Extremely high.
ALTERATION	Sphene alters to **leucoxene**, an aggregate of quartz and rutile to which ilmenite also alters, as follows:

$$2CaTiSiO_4(O,OH,F) \rightarrow 2TiO_2 + 2SiO_2 + 2Ca(O,OH)$$

leucoxene

*BIREFRINGENCE	Extreme, but colours tend to be masked by the body colour and the high relief of the mineral. (This mineral does not change appearance under crossed polars.)
INTERFERENCE FIGURE	$2V$ is small; the small size of most crystals, coupled with the high dispersion of light, means that a good figure is not often obtained.
*TWINNING	Single twins with {100} twin plane common.
OCCURRENCE	It is a common accessory mineral in igneous and metamorphic rocks, particularly plutonic intermediate and acid rocks, such as diorite, granodiorites and granites, and also alkali igneous rocks such as nepheline–syenites. It is common in some skarns and in metamorphic basic schists and gneisses. It is rare in sediments.

Staurolite   Nesosilicate

Staurolite $(Fe,Mg)_2(Al,Fe^{3+})_9O_6Si_4O_{16}(O,OH)_2$ orthorhombic
                                                                 0.471 : 1 : 0.340

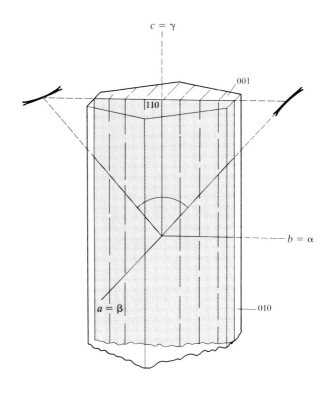

$n_\alpha = 1.739–1.747$
$n_\beta = 1.745–1.753$
$n_\gamma = 1.752–1.761$
$\delta = 0.013–0.014$
$2V_\gamma = 82°–90°$ + ve
OAP is parallel to (100)
$D = 3.74–3.85$ $H = 7\frac{1}{2}$

*COLOUR	Yellow, pale yellow.
*PLEOCHROISM	Always present and distinct in yellows, with α colourless, β pale yellow, and γ a rich golden yellow.
HABIT	Staurolite occurs as squat prisms, usually containing inclusions, particularly of quartz.
CLEAVAGE	{010} moderate.
RELIEF	High.
*ALTERATION	Rare, but may alter to a green ferric chlorite.

147

BIREFRINGENCE	Low, but interference colours are masked by the yellow colour of the mineral.
INTERFERENCE FIGURE	Since $2V$ is very large, a single optic axis is required.
EXTINCTION	Straight on prism edge or cleavage.
OCCURRENCE	Staurolite occurs only in regional metamorphic rocks which are rich in alumina and iron, with probably a high $Fe^{3+} : Fe^{2+}$ ratio. In medium-grade pelitic rocks, staurolite develops from chloritoid and before kyanite, but kyanite and staurolite can coexist. With increasing grade, staurolite breaks down to give kyanite and garnet.

Stilpnomelane   Phyllosilicate

Stilpnomelane $(K,Na,Ca)_{0-1.4}(Fe^{3+},Fe^{2+},Mg,Al,Mn)_{5.9-8.2}[Si_8O_{20}]$-
$(OH)_4(OH,F)_{3.6-8.5}$
?triclinic
$0.557 : 1 : 2.570$
$\alpha = 124°, \beta \approx 96°, \gamma \approx 120°$

$n_\alpha = 1.543-1.634$
$n_\beta = n_\gamma = 1.576-1.745$
$\delta = 0.030-0.110$
$2V_\alpha = 0°$
OAP is parallel to (010)
$D = 2.59-2.96 \qquad H = 3-4$

COLOUR	Various shades of brown; the colour becomes greener with increasing ferrous iron.
PLEOCHROISM	Strongly pleochroic, with α golden yellow, and β and γ dark reddish brown to black. With increasing ferrous iron, β and γ are deep green.
HABIT	Thin platy crystals similar to micas, often in a radiated habit.
RELIEF	Low to moderate.
ALTERATION	Stilpnomelane alters to iron oxides and clays.
BIREFRINGENCE	Extremely high, but masked by the mineral body colour.
INTERFERENCE FIGURE	Stilpnomelane appears uniaxial negative in a basal section.
EXTINCTION	Straight on cleavages.
DISTINGUISHING FEATURES	Stilpnomelane is darker than biotite and is virtually uniaxial.
OCCURRENCE	Stilpnomelane occurs in low-grade metamorphosed Fe- and Mn-rich sedimentary deposits, such as the Lake Superior ironstones, and is also present in some glaucophane schists.

148

Talc   Phyllosilicate

Talc $Mg_6Si_8O_{20}(OH)_4$ monoclinic

$0.577:1:2.068, \beta = 100°00'$

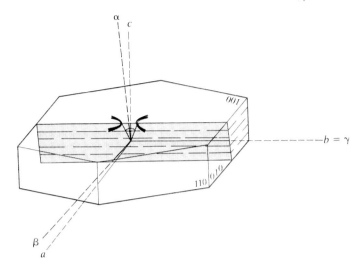

$n_a = 1.539–1.550$
$n_\beta = 1.589–1.594$
$n_\gamma = 1.589–1.600$
$\delta = 0.05$
$2V_a = 0°–30° \ -ve$
OAP is perpendicular to (010)
$D = 2.58–2.83 \qquad H = 1$

COLOUR	Colourless.
HABIT	Tabular crystals or pseudo-hexagonal plates, similar to micas.
*CLEAVAGE	Perfect {001} basal.
RELIEF	Low.
*BIREFRINGENCE	High with interference colours of third order. Basal sections, similar to those of muscovite, give very low first-order greys.
INTERFERENCE FIGURE	Good Bx_a figure with small $2V$ on basal section.
EXTINCTION	Straight.
*OCCURRENCE	Talc occurs by low-grade thermal metamorphism of siliceous dolomites, and by the hydrothermal alteration of ultrabasic rocks, where talc may occur along faults and shear planes. The development of talc is often associated with serpentinization, with the serpentine changing to talc plus magnesite by addition of CO_2.

149

Topaz   Nesosilicate

Topaz $Al_2SiO_4(OH,F)_2$ orthorhombic,
                                                          $0.528:1:0.955$

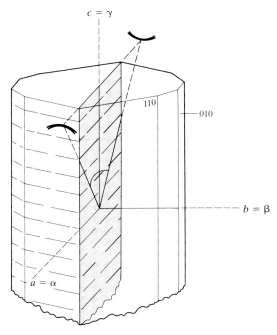

$n_\alpha = 1.606–1.629$
$n_\beta = 1.609–1.631$
$n_\gamma = 1.616–1.638$
$\delta = 0.008–0.011$
$2V_\gamma = 48°–68°$ + ve
OAP is parallel to (010) (a prism section is length-slow)
$D = 3.49–3.57$ $H = 8$

COLOUR   Colourless, but thick sections may be yellowish or pink.
HABIT   Usually occurs as prismatic crystals, subhedral to anhedral.
*CLEAVAGE   {001} perfect.
RELIEF   Moderate.
*INCLUSIONS   Liquid inclusions, either of water or brewsterlinite (originally
                thought to be CO_2) are common.
*ALTERATION   Topaz alters to clay minerals (kaolin) and sericite. One reaction is as
                follows:

$$2Al_2SiO_4(OH,F)_2 + 2H_2O + 2SiO_2 \rightarrow Al_4Si_4O_{10}(OH,F)_8$$
$$\text{topaz} \qquad\qquad\qquad\qquad \text{kaolin}$$

This is a hydrothermal alteration taking place at a late stage, in the
presence of free silica.

150

BIREFRINGENCE	Low.
INTERFERENCE FIGURE	A single optic axis figure is required to obtain the sign.
*OCCURRENCE	Topaz is a mineral found in late-stage acid igneous rocks such as granites, rhyolites and pegmatites, where it can also occur in cavities. It is associated with pneumatolytic action and is a constituent of *greisen*. It occurs with quartz, cassiterite, tourmaline, fluorite and beryl. It has been known to form in metamorphosed bauxite deposits.

Tourmaline group   Cyclosilicate

Tourmaline $Na(Mg,Fe,Mn,Li,Al)_3Al_6Si_6O_{18}(BO_3)_3(OH,F)_4$ trigonal, c/a 0.447

The composition of tourmaline varies greatly, with many types known. Thus **dravite** is the magnesian variety ($NaMg_3$ etc.), **schorl** the ferromanganoan variety ($Na(Fe,Mn)_3$ etc.) and **elbaite** the lithium-bearing type ($Na(Li,Al)_3$ etc.).

$n_o = 1.610–1.630 +$ ⎫
$n_e = 1.635–1.655 +$ ⎬ RI depends on composition
$\delta = 0.021–0.026 +$ ⎭
Uniaxial − ve (all types) (prism sections are length fast)
$D = 2.9–3.2$ $H = 7–7\frac{1}{2}$

*COLOUR	Highly variable; colourless, blue, green or yellow.
*PLEOCHROISM	Elbaite is usually colourless, but the other varieties are pleochroic (pleochroism in crystals of schorl is shown in Plate 2a):

dravite	o	dark brown	e	pale yellow
	o	yellow brown	e	yellow
schorl	o	dark green	e	reddish violet
	o	blue	e	pale green, pale yellow

*HABIT	Tourmaline almost always occurs as large elongate prismatic crystals, often occurring in radiating clusters.
CLEAVAGE	$\{11\bar{2}0\}$ and $\{10\bar{1}1\}$ poor, often appearing perpendicular to the prism zone.
RELIEF	Moderate.
BIREFRINGENCE	Moderate, colours of second order are seen, but are frequently masked if the tourmaline has a strong body colour (see Plate 2b).
TWINNING	Rare.
ZONING	Colour zoning may occur.
OCCURRENCE	Tourmaline typically occurs in granite pegmatites, pneumatolytic veins and some granites as the schorl–elbaite type. In the pneumatolytic stage of alteration, tourmaline may form by boron introduction, and the rock *luxullianite* forms in this environment. In

151

pneumatolytic igneous assemblages, tourmaline is associated with topaz, spodumene, cassiterite, fluorite and apatite. In metamorphic rocks (especially metamorphosed limestones) and metasomatic rocks, the dravite type of tourmaline occurs; and dravites have been recorded from basic igneous rocks. Tourmalines have been found as detrital minerals in sedimentary rocks, and as authigenic minerals in some limestones.

Vesuvianite   Sorosilicate

Vesuvianite (or idiocrase)

$Ca_{10}(Mg,Fe)_2Al_4Si_9O_{34}(OH.F)_4$ tetragonal, c/a 0.757

$n_o = 1.708–1.752$

$n_e = 1.700–1.746$

$\delta = 0.001–0.008$

Uniaxial $-$ ve (a prism section is length fast)

$D = 3.33–3.43$ $H = 6–7$

COLOUR Colourless, pale yellow or pale brown.

HABIT Prismatic crystals usually occur, but in general crystals are sub-hedral with only a few faces present.

CLEAVAGE {110} and {100} poor.

RELIEF High.

BIREFRINGENCE Low, greys of first order.

DISTINGUISHING FEATURES Vesuvianite is a difficult mineral to recognize; in relief and birefringence it resembles zoisite (see "Epidote group"). However, it usually occurs as large mineral grains, and its occurrence and mineral associations are most important.

*OCCURRENCE Vesuvianite or idocrase occurs in thermally metamorphosed limestones and in skarns. It is associated with grossular (Ca-bearing) garnet, diopside and wollastonite. It has been found in nepheline–syenites and in veins in basic igneous rocks.

Zeolite group   Tektosilicates

The zeolites have the general formula $(Na_2,K_2,Ca,Ba)[(Si,Al)_2]_n$. yH_2O. They occur in amygdales and cavities of basic igneous volcanic rocks, where they have been deposited by late-stage hydrothermal solutions passing through the rocks. The most important members are:

natrolite Na_2 $[Al_2Si_3O_{10}]. 2H_2O$

mesolite $Na_2Ca_2[Al_2Si_3O_{10}]_3.8H_2O$

scolecite $Ca [Al_2Si_3O_{10}]. 3H_2O$

thomsonite $NaCa_2[(Al,Si)_5O_{10}]. 6H_2O$

The above zeolites are all fibrous minerals. The zeolites listed below crystallize in various types of crystal habit – tabular, platy or prismatic:

chabazite	Ca	$[Al_2Si_4O_{12}].6H_2O$
heulandite	Na_2Ca	$[Al_2Si_7O_{18}].6H_2O$
stilbite	$Na_2CaK_2[Al_2Si_7O_{18}].7H_2O$	
laumontite	Ca	$[Al_2Si_4O_{12}].4H_2O$

Analcime is also a zeolite, and should be included in this group of minerals. However, because analcime is closely associated with the *feldspathoids* it has been included with that mineral group (see p. 97). The zeolites are widely used as indicator minerals in thick laval piles, such as *ocean floor basalts*, to determine the temperature and depth of burial. A typical sequence of zeolite minerals from a recent Icelandic lava pile is (from top to bottom):

> zeolite-free zone
> chabazite–thomsonite
> analcime (plus natrolite)
> mesolite–scolecite

Although other lava piles may show slight variations in the zeolites present, the zones described above generally occur. Natrolite and most other zeolites are colourless in thin section, with RIs very much lower than the cement. They mostly belong to orthorhombic or monoclinic crystal systems (natrolite is orthorhombic), with either straight extinction or a small extinction angle. Natrolite is length-slow. Their $2V$ is usually large, and either positive or negative, and their birefringence is variable but low. Their occurrence in vesicles and amygdales is the most reliable indicator for identification. X-ray diffraction techniques are required for the positive identification of zeolite type. The main optical properties of the zeolites are as follows:

*COLOUR — Colourless.

HABIT — Apart from analcime (see 'Feldspathoid family'), most zeolites are elongate fibrous or platy, often occupying cavities or amygdales in extrusive igneous rocks.

CLEAVAGE — Variable, depending upon the crystal system. Most fibrous varieties possess at least one prismatic cleavage.

*RELIEF — Low to moderate; RI is less than 1.54 for all minerals.

ALTERATION — Rare, but a few zeolites will alter to clay minerals.

*BIREFRINGENCE — Generally low to very low. A very few zeolites may show first-order yellow.

INTERFERENCE FIGURE — Variable.

EXTINCTION — All fibrous varieties have straight extinction on the prism edge, except for scolecite. Platy varieties usually possess inclined extinction.

TWINNING | Simple twinning is common in mesolite, laumonite, chabazite and stilbite. Multiple twinning is common in scolecite, phillipsite and harmotome.

Zircon　　　　　　　　　　　　　　　　　　　　　　　Nesosilicate

Zircon　ZrSiO$_4$　　　　　　　　　　　　　　　tetragonal, c/a 0.891

n_o = 1.923–1.960
n_e = 1.968–2.015
δ = 0.042–0.065
Uniaxial + ve (a prism section is length slow)
D = 4.6–4.7　　H = $7\frac{1}{2}$

COLOUR | Colourless pale brown.
*HABIT | Very small, squat, square prisms occur with terminal faces. Zircons are usually found as euhedral crystals.
CLEAVAGE | {110} imperfect; {111} poor.
RELIEF | Extremely high.
ALTERATION | None.
*BIREFRINGENCE | Very high: a prismatic crystal section will show third- or fourth-order interference colours.
TWINNING | Rare.
ZONING | May be present due to outer metamict zones on an unaltered core.
DISTINGUISHING FEATURES | Tiny euhedral crystals in alkaline or acid plutonic igneous rocks are usually zircon. Sphene is pale brown, usually with a diamond-shaped cross section, and is biaxial positive. Monazite is biaxial positive. Cassiterite and rutile are coloured minerals.
*OCCURRENCE | An accessory mineral found in all igneous rocks, but essentially in intermediate to acid varieties, where it is associated with biotite crystals. Haloes frequently occur in the biotite surrounding minute zircon crystals (due to radioactive decay of U and Th damaging the biotite structure by β-particle bombardment). Zircon is most commonly found in plutonic igneous rocks; particularly granites, granodiorites, diorites, syenites, nepheline–syenites and pegmatite veins. Zircon also occurs as a detrital mineral in sediments, and will survive many metamorphic and melting events.

3 The non-silicates

3.1 Introduction

Minerals which are not silicates have been grouped together in this chapter for the description of their properties. However, unlike the silicates, the crystal structures and chemical variation of members of the group are not easily related to mineralogical properties and the mode of occurrence. Even subdivision of the group into transparent and opaque minerals is impractical, since closely related minerals, and even compositional varieties of the same mineral, may vary in opacity. For example, sphalerite is transparent when it is pure zinc sulphide, but it becomes progressively more opaque with increasing iron substitution of zinc (see Plate 4a).

The non-silicates can usually be regarded as accessory minerals in most rocks, yet they are major components in some rock types, e.g. halides in evaporites, sulphides in massive sulphide deposits and carbonates in limestones.

Minerals of the following non-silicate groups appear in this chapter in alphabetical order: arsenides (As^{2-}), carbonates (CO_3^{2-}), halides (Cl^-, F^-), hydroxides (OH^-), native elements, oxides (O^{2-}), phosphates (PO_4^{3-}), sulphates (SO_4^{2-}), sulphides (S^{2-}, S_2^{2-}) and tungstates (WO_4^{2-}). Within each group, the minerals are described in alphabetical order. The relationship of the optical and physical properties to the chemical composition and structure is outlined only for the four major groups: carbonates, halides, oxides and sulphides.

In this chapter, where appropriate, thin-section information is as described in Section 1.4 and presented for the silicates in Chapter 2. The polished-section information, for observations using reflected light, is as described in Section 1.7.

3.2 Arsenides

Niccolite   NiAs

The name recommended by the International Mineralogical Association is nickeline. Niccolite may contain some Fe or Co.

155

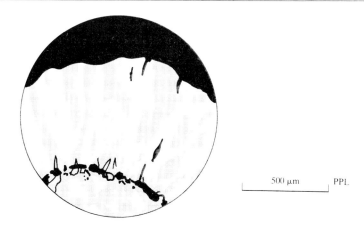

500 μm   PPL

Figure 3.1   Radiating intergrowth of niccolite in different crystallographic orientations.

Crystals   Niccolite is hexagonal, $a:c = 1:1.3972$. Crystals are rare. It is usually massive, reniform with columnar structure. Repeated twinning occurs on $\{10\bar{1}1\}$. There is no cleavage. $D = 7.8$. $H = 4\frac{1}{2}$.

Polished   Niccolite is pinkish or orange white with a pronounced pleoch-
section   roism, with $R_o = 52\%$ (lighter, orange or yellowish) and $R_e = 48\%$ (darker, pinkish). The reflectance is similar to pyrite. Anisotropy is very strong, the tints being bright bluish and greenish greys.

   Niccolite usually occurs in xenomorphic or concentric, botryoidal masses with other Co + Ni + As + S minerals. Grains are often cataclased. Growth zonation is common, and botryoidal masses often contain radiating intergrown irregular lamellae (Fig. 3.1). VHN = 328–455.

Occurrence   Niccolite occurs in Ni + Co + Ag + As + U deposits, which are probably low-temperature hydrothermal veins and replacements. Such deposits are often associated with basic igneous rocks and organic-rich sedimentary rocks.

Distinguishing Compared with niccolite, marcasite is whiter, and arsenopyrite is
features   whiter and has a weaker anisotropy.
Note   Niccolite alters to green annabergite.

3.3   Carbonates

The carbonates, of which the best known example is calcite $CaCO_3$, contain a discrete $(CO_3)^{2-}$ radical that may be considered as a single anion in the structure, but is in fact a trigonal planar complex. This complex, with carbon in the centre of an equilateral triangle formed by three oxygens, is shown in the carbonate structure in Figure 3.2.

Figure 3.2
The structure
of calcite
CaCO₃.

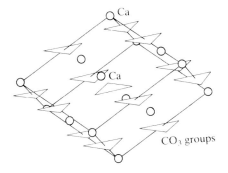

There are relatively few common carbonates of rock-forming significance, and most can be considered as secondary or replacive minerals forming on alteration of metal-bearing precursor minerals, e.g. cerussite PbCO₃ after galena PbS. Some secondary carbonates contain structural water, e.g. malachite $Cu_2CO_3(OH)_2$ after chalcopyrite $CuFeS_2$.

The triangular nature of the $(CO_3)^{2-}$ radical dominates the structure of the carbonates and results in trigonal (rhombohedral) or orthorhombic (pseudo-hexagonal) symmetry. The critical factor controlling the type of symmetry is the radius of the dominant metallic cation; for elements such as Mn, Fe and Mg with a radius less than about 1.0 Å the carbonates are trigonal, but for elements such as Ba, Sr and Pb with large radii the carbonates are orthorhombic. Calcium lies close in radius value to the critical size, and this explains the existence of CaCO₃ as two minerals, calcite (trigonal) and aragonite (orthorhombic). Although aragonite is considered to be a high-pressure polymorph of CaCO₃, it can grow at low pressures provided that the solution chemistry is correct. However, it is metastable and it usually inverts to calcite during recrystallization processes such as diagenesis.

Figure 3.3
Carbonates in the
$CaCO_3$–$MgCO_3$–$FeCO_3$
system.

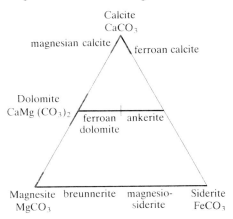

Table 3.1 Optical properties of the common carbonates.

Trigonal structures (uniaxial)			n_o	n_e	Optic sign
calcite $CaCO_3$			1.658	1.486	− ve
dolomite $CaMg(CO_3)_2$			1.679	1.500	− ve
siderite $FeCO_3$			1.875	1.635	− ve
rhodochrosite $MnCO_3$			1.816	1.597	− ve
magnesite $MgCO_3$			1.700	1.509	− ve

Orthorhombic structures (biaxial)	n_α	n_β	n_γ	$2V$ (degrees)	Optic sign
strontianite $SrCO_3$	1.518	1.665	1.667	8	− ve
witherite $BaCO_3$	1.529	1.676	1.677	16	− ve
aragonite $CaCO_3$	1.530	1.680	1.685	18	− ve
cerussite $PbCO_3$	1.803	2.074	2.076	9	− ve

Chemical substitution is quite extensive in the common carbonates, e.g. manganoan calcite $(Ca,Mn)CO_3$ and magnesian siderite $(Fe,Mg)CO_3$. The range in composition of carbonates belonging to the $CaCO_3$–$MgCO_3$–$FeCO_3$ system is illustrated in the triangular diagram of Figure 3.3. The compositional ranges are only approximate, but are given to emphasize the fact that minerals such as dolomite can differ quite significantly from the stoichiometric end-member composition. Most dolomites are ferroan, and iron-rich dolomite is called "ankerite". Manganese is the main additional element that is found substituting into these carbonates, and kutnahorite is the manganese-rich variety of dolomite. Chemical staining, cathodoluminescence microscopy or electron-beam microanalysis is recommended to characterize this common group of minerals, but optical studies can provide some useful preliminary information.

The carbonate minerals have very large birefringences (Table 3.1) and they usually have well developed cleavages and multiple twinning (Fig. 3.4). The large birefringence is due to the planar triangular $(CO_3)^{2-}$ radicals, which are oriented normal to the c axis. It is usually easy to determine in thin section that a mineral is a carbonate, but identification of the particular carbonate is difficult. Clues include the grain shape, with dolomite often being rhomb-shaped while calcite tends to be anhedral. Alteration, due to oxidation, of iron-bearing carbonates leads to a penetrative yellowish or reddish-brown staining, whereas manganese-bearing carbonates yield a black alteration product. However, care must be exercised, since pure calcite will react with metal-bearing acidic water to produce a

Figure 3.4 Typical polycrystalline calcite as found in marble. Note the high but variable relief and the multiple twinning (PPL).

Figure 3.5 Calcite (grey patchwork) intergrown with sphalerite (light grey). The patchwork effect results from the bireflectance of calcite, and the grains vary in brightness as the stage is rotated (R-PPL).

coloured alteration product. Hand-specimen colours, such as the pink colour of many dolomites, need not be observed in thin section; when it is well crystallized, even malachite can show little of its distinctive green colour. In polished section, the carbonates have low reflectance values, but have distinct bireflectances due to the large birefringence (Fig. 3.5). It is useful to remember that the reflectance depends on the refractive index, so that calcite and dolomite in intergrowths can usually be more readily distinguished on the basis of reflectance in polished section than on the basis of relief in thin section, provided that care is taken to compare only the highest (o-ray) reflectances of the grains (see Appendix E).

Aragonite $CaCO_3$ orthorhombic
                                                              $0.6228:1:0.7204$

$n_\alpha = 1.530$
$n_\beta = 1.680$
$n_\gamma = 1.685$
$\delta = 0.155$
$2V_\alpha = 18°$ − ve (crystals are length fast)
OAP is parallel to (100)
$D = 2.94$ $H = 3\frac{1}{2}$

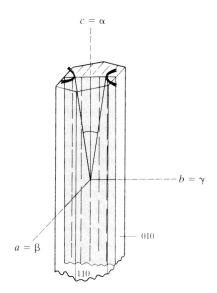

COLOUR	Colourless.
*HABIT	Thin prismatic or occasionally fibrous crystals occur as, for example, in shell structures.
CLEAVAGE	{010} prismatic cleavage imperfect.
RELIEF	Low to moderate, but variable with optic orientation, as for calcite.

The minimum RI is parallel to the c axis (i.e. parallel to prism length).

*BIREFRINGENCE   Extremely high, similar to calcite.

INTERFERENCE FIGURE   Difficult to obtain because of the crystal size, but a good Bx_a figure may be seen on a basal section ($2V$ very small).

EXTINCTION   Straight on cleavage or prism edge.

TWINNING   Common; lamellar twins on {110}, parallel to the c axis. Repeated twinning is also common.

OCCURRENCE   Aragonite is less common than calcite. Many invertebrates build their shells of aragonite, which gradually changes to calcite on diagenesis. Thus pre-Mesozoic fossil shells will inevitably consist of calcite. Aragonite occurs as a secondary mineral, often in association with zeolites, in cavities in volcanic rocks. It is a widespread metamorphic mineral in glaucophane–schist facies metamorphic rocks in which deep burial produces aragonite as the stable carbonate at $\sim 300°C$ and a pressure of 6–10 kb. Aragonite inversion to calcite may occur as the rock attains normal P–T conditions.

Calcite $CaCO_3$ trigonal, c/a 0.8550

$n_o = 1.658$

$n_e = 1.486$

$\delta = 0.172$

Uniaxial $-$ ve

$D = 2.715$ $H = 3$

COLOUR   Colourless.

HABIT   Often as shapeless grains (anhedral); occasional rhombohedral outline seen in sedimentary limestones.

CLEAVAGE   Perfect {10$\bar{1}$1} rhombohedral cleavage – three cleavage traces seen in some sections.

*RELIEF   Moderate, with extreme variation because of large birefringence. Note that the refractive indices cover a range of values which 'bracket' 1.54. The crystal is said to "twinkle" during rotation. Prismatic crystals parallel to the c axis are length-fast.

*BIREFRINGENCE   Extremely high, with pale pinks and greens of fourth order and higher.

INTERFERENCE FIGURE   Because of the large birefringence, grains show moderate-order interference colours even when the optic axis is near vertical, and these can be used to obtain a uniaxial interference figure.

TWINNING   {01$\bar{1}$2} common, appearing as multiple twins; {0001} common, simple twin plane.

OCCURRENCE   One of the most common non-silicate minerals. It is a principal constituent of sedimentary limestones, occurring as carbonate shell material, as fine precipitates and as clastic material. Shells are generally composed of calcite or aragonite. Aragonite usually occurs as the initial carbonate precipitate, but it eventually re-crystallizes to calcite.

On metamorphism, pure calcitic limestone recrystallizes to marble, in which the calcite grains are welded together in a mosaic; in impure limestones the calcite combines with the impurities present to give new minerals, the type of mineral depending upon the temperature and CO_2 pressure. The reaction

$$\text{calcite} + \text{silica} \rightarrow \text{wollastonite (CaSiO}_3) + \text{CO}_2$$

occurs at $\approx 600°C$ at low pressures, but the same reaction occurs at over 800°C as the pressure increases. Calcite can also occur with calc-silicate minerals such as diopside, garnet (Ca-rich) and idocrase (vesuvianite) in metamorphic rocks.

Calcite may occur in vugs or cavities in igneous rocks, being deposited by late-stage hydrothermal solutions. In hydrothermal veins, calcite is a common gangue mineral, often being found with fluorite, barite or quartz and in association with the sulphide ore minerals. Calcite may occur as a primary crystallizing mineral in some igneous rocks, particularly carbonatites and some nepheline–syenites. Calcite may also occur as a secondary mineral on alter-ation of ferromagnesian minerals by late-stage hydrothermal solu-tions in which CO_2 is present.

| **Dolomite** | $CaMg(CO_3)_2$ | trigonal, c/a 0.8235 (dol) |
| | | 0.835 (ank) |

Dolomite is one end member of a mineral series between dolomite and **ankerite** $Ca(Mg,Fe)(CO_3)_2$, the iron-rich end member. Fe can replace Mg in dolomite, but when Mg < Fe the mineral is called ankerite.

	Dolomite	Ankerite	
n_o	1.679	1.690–1.750	⎱ RIS increase with
n_e	1.500	1.510–1.548	⎰ increasing iron
δ	0.179	0.180–0.202	
	Uniaxial − ve	Uniaxial − ve	
	$D = 2.86$ $H = 3\frac{1}{2}$	$D = 3.04$ $H = 4$	

COLOUR Colourless.

*HABIT Usually subhedral, but dolomitization of limestones often leads to euhedral crystals, occurring as rhombohedra with curved faces (baroque dolomite).

CLEAVAGE Perfect $\{10\bar{1}1\}$ rhombohedral, as calcite.

RELIEF Low to moderate (variable with optic orientation).

*BIREFRINGENCE Extremely high, even higher than calcite (almost colourless, but a slight iridescence gives an indication of the extreme birefringence).

162

*TWINNING	Similar to calcite, i.e. multiple on $\{02\bar{2}1\}$. The twin lamellae show birefringence of a lower order than the crystal.
ZONING	Commonly encountered, Fe^{2+} substitution of Mg^{2+} (ankerite).
OCCURRENCE	Note that dolomite is also a name given to rock consisting mainly of dolomite. Dolomite occasionally occurs as a primary mineral in sedimentary rocks and is often associated with evaporite deposits. As a secondary mineral, dolomite is formed during dolomitization of a limestone shortly after deposition, and before consolidation. Another type of dolomitization occurs after consolidation of a limestone if Mg-rich solutions enter the rock. Dolomite is currently forming in certain saline lakes. The formation of primary and secondary dolomite may be due to the marine environment changing from deep to shallow water, with increasing salinity. Dolomite can occur as a gangue mineral with fluorite, barite, calcite, quartz or siderite in association with lead and zinc sulphides. Dolomite rock is commonly associated with serpentines and other ultramafic rocks, and it is common in ophiolite suites. During metamorphism dolomitic marbles may crystallize from dolomitic limestones. At higher grades of metamorphism the dolomite eventually breaks down to give periclase MgO, with brucite, $Mg(OH)_2$ forming on hydration.

Siderite $FeCO_3$ trigonal, c/a 0.819

$n_o = 1.782$
$n_e = 1.575$
$\delta = 0.207$
Uniaxial $-$ ve
$D = 3.5$ $H = 4\frac{1}{2}$

COLOUR	Colourless to pale brown or pale yellow.
HABIT	Euhedral crystals are common, and siderite is often found as aggregates of crystals in oolitic structures.
	All other properties are similar to those of calcite (note the extreme birefringence).
OCCURRENCE	Common in ironstone nodules in Carboniferous argillaceous rocks, and also in the Jurassic ironstones of central England. In Raasay in the Inner Hebrides, siderite is associated with chamosite.
	Siderite is found in veins with other gangue minerals and metallic ores.

3.4 Halides

Halides are ionic minerals that consist essentially of metallic cations and halogen anions. The common examples are normal anhydrous halides, which have simple chemical compositions and structures,

163

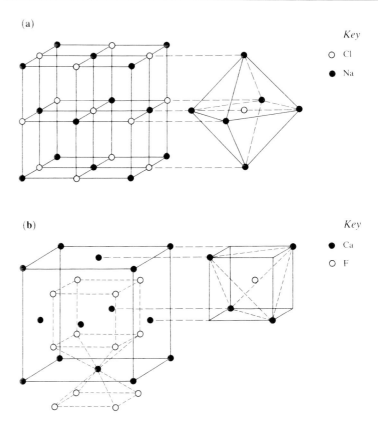

(a)

Key

○ Cl

● Na

(b)

Key

● Ca

○ F

Figure 3.6 (a) The crystal structure of sodium chloride, showing (right) the octahedral arrangement of six sodium ions around one chloride ion. (b) Fluorite structure showing (right) the tetrahedral arrangement of four calcium ions around one fluoride ion, and (left) the cubic arrangement of eight fluoride ions around one calcium ion.

but there is a host of related oxyhalides, hydroxyhalides and complex-containing halides. One of the aluminofluorides, cryolite Na_3AlF_6, is worthy of mention, but the others are of little significance as rock-forming minerals. The structures of fluorite CaF_2 and halite (rock salt) NaCl are illustrated in Figure 3.6. Sylvite KCl has the halite structure.

The simple halides are typical ionic compounds, there being a large difference in the ionization potentials of the metal and halogen atoms: they are typical transparent insulators. Both halite and fluorite have low refractive indices and are therefore quite dark (i.e. they have a small reflectance) in polished section.

Fluorite CaF_2 cubic
$n = 1.433–1.435$ (variation due to substitution of Y for Ca)
$D = 3.18$ $H = 4$

COLOUR   Colourless, very pale green, pale blue, yellow or violet.
HABIT   Aggregates of crystals, often with perfect cubic {100} form.
CLEAVAGE   {111} perfect, giving a triangular pattern.
*RELIEF   Moderate, but note that n is less than 1.54.
TWINNING   Interpenetrant on {111}, but not seen in thin section.
OCCURRENCE A late-stage mineral in granites and other acid rocks; common in greisen. In pegmatites, and many alkaline igneous rocks such as nepheline–syenites, fluorite crystallizes at low temperatures (around 500 °C). In late-stage pneumatolytic deposits, fluorite occurs with cassiterite, topaz, apatite and lepidolite, whereas in hydrothermal veins fluorite occurs with calcite, quartz, barite and sulphides.

Fluorite is occasionally found as the cementing matrix in sandstones and may occur in geodes within limestones. Blue John is coarse nodular purple fluorite, with a concentric layered structure.

Halite   NaCl   cubic
(rock salt)

$n = 1.544$
$D = 2.16$ $H = 2\frac{1}{2}$

Halite is colourless rock salt with a perfect {100} cubic cleavage. It occurs in salt domes and in evaporites, where it is a late-precipitating salt.

Note: Special sectioning techniques are needed to preserve this mineral in thin sections.

3.5   Hydroxides

Gibbsite (Hydrargillite)
$Al(OH)_3$ monoclinic
$1.7043:1:1.9170, \beta = 94° 34'$

$n_\alpha = 1.568–1.580$
$n_\beta = 1.568–1.580$
$n_\gamma = 1.587–1.600$
$\delta \approx 0.019$
$2V_\gamma = 0°–40°$ + ve
OAP is perpendicular to section (010)
$D = 2.38–2.42$ $H = 2\frac{1}{2}–3\frac{1}{2}$

COLOUR Colourless, pale brown.

*HABIT Tiny tabular pseudo-hexagonal crystals on {001} common; some-times occurring as aggregates.

*CLEAVAGE {001} basal cleavage perfect.

RELIEF Low to moderate.

ALTERATION Gibbsite forms from the alteration of feldspar, nepheline and other aluminous minerals. It may form from the hydration of boehmite. It is stable under normal atmospheric conditions, but may change to kaolin by silicification.

BIREFRINGENCE Interference colours are white or yellow of first order.

*INTERFERENCE FIGURE Gibbsite is biaxial positive with a small $2V$. Cleavage plates (i.e. basal sections) yield off-centre Bx_a figures.

EXTINCTION ANGLE An (010) section shows oblique extinction on the basal cleavage with an angle of fast $(\beta)\hat{\ }$cleavage $\approx 26°$.

DISTINGUISHING FEATURES Gibbsite can be confused with other clay minerals, but most of these are negative, and with straight extinction. Diaspore and boehmite have higher relief, larger $2V$ and straight extinction.

OCCURRENCE Gibbsite forms from intense chemical weathering of aluminous minerals, appearing in lateritic soils and bauxites, where it may be the dominant constituent, in association with diaspore, boehmite and ferric oxides. Gibbsite may occasionally appear as an hydro-thermal mineral in low-temperature veins in Al-rich igneous rocks.

Brucite $Mg(OH)_2$ trigonal, c/a 1.5154

$n_o = 1.560–1.590$
$n_e = 1.580–1.600$
$\delta = 0.012–0.020$
Uniaxial + ve (length fast)
$D = 2.4$ $H = 2\frac{1}{2}$

COLOUR Colourless.

HABIT Occurs as fine aggregates, or fibrous whorls, in metamorphosed impure limestones.

*CLEAVAGE Perfect basal {0001}.

RELIEF Low, just greater than 1.54.

ALTERATION Brucite forms from **periclase** MgO by addition of H_2O during thermal metamorphism. It alters to **hydromagnesite** readily by reac-tion with carbon dioxide:

166

$$5Mg(OH)_2 + 4CO_2 \rightarrow Mg_5(OH)_2(CO_3)_4.4H_2O$$
hydromagnesite

*BIREFRINGENCE	Low, first-order colours, but often shows anomalous interference colours (deep blue), rather similar to chlorite.
TWINNING	None.
OCCURRENCE	Brucite occurs in thermally metamorphosed dolomites, and dolomitic limestones. It can occur in low-temperature hydrothermal veins, associated with serpentinites and chlorite schists.
DISTINGUISHING FEATURES	Brucite has anomalous birefringence and is uniaxial positive. Micas and talc have higher birefringence and are optically negative, as is gypsum. Apatite and serpentine are very similar optically, but serpentine is usually greenish and both are *always* length-slow.

Limonite
Goethite
Lepidocrocite

$FeO.OH.nH_2O$

Limonite is brown earthy material consisting of geothite ± lepidocrocite with absorbed water.

Crystals	Goethite a-FeO.OH is orthorhombic, $a:b:c = 0.4593:1:0.3034$. Lepidocrocite γ-FeO.OH is orthorhombic, $a:b:c = 0.309:1:0.245$. Both minerals occur as flakes or blades flattened (010) or as fibres elongated [100]. There is perfect cleavage on {100}, {010} and {001}. $D = 4.28$, $H = 6$ (geothite); $D = 4.0$, $H = 6$ (lepidocrocite).
Thin section	Poorly crystallized limonite appears reddish brown and isotropic. Goethite is yellowish to brownish, pleochroic with absorption $a < \beta < \gamma$, and has a small negative $2V$. Lepidocrocite is yellow to brownish red, strongly pleochroic with absorption $a < \beta < \gamma$, and has a negative $2V = 83°$.
Polished section	Poorly crystallized limonite is bluish grey, with $R = 16$–19%; anisotropy is strong in bluish greys, and deep red to brown internal reflections are typical. Goethite is grey with $R \approx 17\%$; anisotropy is distinct in shades of grey. Lepidocrocite is grey with $R \approx 11$–18%; anisotropy is very strong in slightly bluish light greys. Internal reflections are deep red to brown in both minerals.
	Limonite is often inhomogeneous, varying in colour or porosity. Goethite is usually colloform and botryoidal, whereas lepidocrocite is usually better crystallized. VHN = 400 (goethite), 690–782 (lepidocrocite).
Occurrence	Limonite is very common as a weathering product after iron-bearing minerals, especially iron carbonates and iron sulphides (Fig. 3.7). It is associated with other hydroxides and oxides in various type of gossans.

167

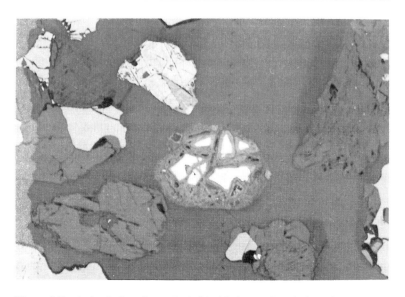

Figure 3.7 A detrital pyrite grain (white) being replaced along fractures by goethite (grey) (R-PPL).

Distinguishing features Compared with limonite, hematite is brighter, harder and has only scarce internal reflections, and sphalerite is isotropic and usually differs texturally.

Diaspore–boehmite series

Diaspore α-AlO(OH) The formulae can be written as α- or γ-HAlO$_2$.
Boehmite γ-AlO(OH)
orthorhombic
0.4664 : 1 : 0.3017

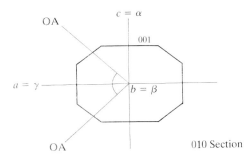

	Diaspore	Boehmite
n_α	1.682–1.706	1.64–1.65
n_β	1.705–1.725	1.65–1.66
n_γ	1.730–1.752	1.65–1.67
δ	0.052–0.046	≈ 0.015

$2V_\gamma = 84°–86° + $ ve $\quad \approx 80° + $ ve

OAP in both minerals is parallel to (010)

$D = 3.3–3.5 \quad 3.01–3.06$

$H = 6\frac{1}{2}–7 \quad\quad 3\frac{1}{2}–4$

The descriptions below are for diaspore, with any variations for boehmite also mentioned.

COLOUR Colourless.

HABIT Both minerals are platy on {010}; but fibrous aggregates are common.

CLEAVAGE Both minerals have a perfect cleavage on {010}. Boehmite also possesses a good cleavage on {100}, and diaspore has an imperfect one on {110}.

RELIEF Moderate to high.

*ALTERATION Both polymorphs are alteration products from aluminous minerals such as nepheline by processes of intense weathering. Silicification of diaspore or boehmite will yield kaolin.

*BIREFRINGENCE Diaspore has strong birefringence of vivid third-order colours seen on an (010) section. Boehmite shows only low first-order colours on a similar section.

INTERFERENCE FIGURE Both minerals are biaxial positive with large $2V$s, so a single isogyre is needed to obtain the sign.

OCCURRENCE Both minerals (and also gibbsite) are components of *bauxite*, together with ferric hydroxides and quartz. They also occur in laterite soils and argillaceous rocks. Boehmite may be metastable.

Diaspore occurs in metamorphosed emery deposits, together with corundum, magnetite, spinel, rutile etc. It also may occur in chlorite schists with chloritoid, kyanite and other aluminous minerals. Diaspore may occur as a product of the hydrothermal alteration of aluminous minerals in association with **alunite** $KAl_3(SO_4)_2(OH)_6$, pyrophyllite, kaolin and topaz.

3.6 Native elements

Copper Cu

Copper may contain As, Ag or Bi.

Crystals Copper is cubic. $D = 8.95$, $H = 3$.

Polished section Copper is bright metallic pink but tarnishes and darkens rapidly. $R \approx 75\%$. It is isotropic, but with incomplete extinction, and fine scratches may cause false anisotropy. VHN = 80–100.

169

Occurrence   Copper occurs as small flakes, granular aggregates, porous masses or dendrites. Zonal texture is not uncommon, and lamellar twinning may be revealed by etching.

It is associated with cuprite Cu_2O and $Cu + Fe + S$ minerals, often in deposits associated with basic extrusives. Copper is common in the oxidation zone, where it results from the oxidation of copper sulphides.

Distinguishing features   Compared with copper, gold is brighter and coloured yellow or white.

Gold   Au

Gold may contain Ag, Cu, Pd or Rh.

Crystals   Gold is cubic and it occurs as cubic, dodecahedral or octahedral crystals, but repeated twinning on {111} often gives reticulated and dendritic aggregates. $D = 19.3$, $H = 2$.

Polished section   Gold is bright yellow. Argentiferous gold is whiter and cupriferous gold is pinker. $R = 76\%$, making gold much brighter than pyrite and chalcopyrite. It is isotropic, but with incomplete extinction, when a greenish colour is observed. Gold does not tarnish, but large grains scratch easily and may be difficult to polish.

Gold occurs as irregular grains, blebs or veinlets, often in sulphides (e.g. pyrite and arsenopyrite). The various varieties of gold are often intergrown with each other or with $Au + Bi + Te$ and $Sb + As$-containing minerals. Gold occurs as very fine coatings which can easily be lost on polishing. VHN = 30–35.

Occurrence   Gold is found in hydrothermal deposits, often associated with igneous rocks; in placer deposits, where it appears to be chemically mobile, resulting in nugget growth; and in auriferous quartz veins. It seems to be present throughout the temperature range of vein mineralization. Gold often occurs as very small grains, even in economic gold deposits.

Distinguishing features   Compared with gold, chalcopyrite is less yellow, darker and weakly anisotropic.

Notes   Electrum (Au, Ag) contains 30–45% Ag. It is brighter ($R \approx 85\%$) and whiter than pure gold.

Graphite   C

Crystals   Graphite is hexagonal, $a:c = 1:1.27522$. The layered structure results in a perfect {0001} cleavage. Crystals are hexagonal tablets {0001}. $D = 2.1$, $H = 1$.

Thin section   In very thin flakes graphite is deep blue and uniaxial negative.

Polished section   Graphite is brownish grey, with a marked pleochroism from $R_o = 26\%$ (grey) to $R_e = 6\%$ (dark brownish grey). It appears slightly brighter than gangue minerals. The anisotropy is strong, in

170

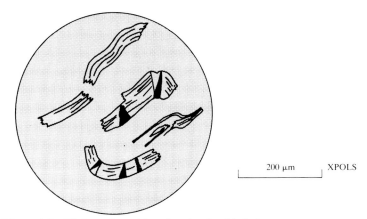

200 μm   XPOLS

Figure 3.8   Flakes of graphite, showing buckled cleavage.

yellowish greys. Extinction is parallel to the cleavage (corresponding to the grey of R_o in PPL), but deformation commonly results in undulose extinction.

Graphite occurs as flakes, tabular crystals, aggregates or botryoidal masses. Flakes are sometimes very long and broken or buckled. The cleavage is usually evident and often deformed (Fig. 3.8). In fact, graphite is rather difficult to polish, and surfaces of large grains are often damaged. VHN = 10.

Occurrence Graphite is common in metasediments, where it forms from organic material; when abundant, a graphitic schist results. Such graphite is indicative of reducing conditions, and pyrite is usually also present. Graphite also occurs in vein-like deposits and large masses, some of which are of uncertain origin.

Distinguishing features Compared with graphite, molybdenite is texturally similar but much brighter.

Note Small flakes of graphite in metamorphic rocks are much more evident using oil immersion.

Silver Ag
Silver may contain minor amounts of Au, Hg, As, Sb, Pt, Ni, Pb or Fe.

Crystals Silver is cubic. $D = 10.5$, $H = 2\frac{1}{2}$.

Polished section Silver is white but it soon tarnishes. With $R \approx 95\%$, it is much brighter than the common ore minerals. It is isotropic, but false anisotropy may result from fine polishing scratches.

Silver occurs in dendritic or irregular masses and as inclusions, often in silver-bearing sulphides or sulphur-poor minerals. VHN = 46–118.

Occurrence Silver is found with Co + Ni + Fe arsenides, usually associated with basic igneous rocks. It also occurs in the oxidized zones of galena-

171

bearing veins. Many veins recorded as silver veins are in fact argentiferous galena veins, the silver being produced as a byproduct of lead recovery. Silver is associated with native copper, and often with carbonate.

3.7 Oxides

Oxides are minerals that contain one or more metals and oxygen; quartz SiO_2 is usually excluded from the group. The reader is referred to Rumble (1976) for a review of the oxides. As most silicates in igneous and metamorphic rocks consist essentially of silica plus metal oxides, free oxides can be considered to form if metal oxides are present surplus to the needs of silicates. Alternatively, they may form if rocks are silica deficient, as is usually the case when periclase MgO and corundum Al_2O_3 are found; or they may form if the metal is "inappropriate" for a silicate structure, e.g. cassiterite SnO_2. The two following groups, the iron–titanium oxides and the spinels, overlap to a certain degree, but will be outlined briefly because they contain the most common oxides. Note that the oxides are described in alphabetic order.

Iron–titanium oxides
Minerals with chemical compositions primarily of iron, titanium and oxygen occur widely in rocks of all types. Their identification is important because much can be learnt about the crystallization history of the host rock. Rumble (1976) states: "The oxide minerals are of great value in deducing the conditions of metamorphism; indeed, their value is out of all proportion to their modal abundance in typical rocks, for they simultaneously record information on both the ambient temperature and the chemical potential of oxygen during metamorphism." The same statement may be applied to igneous rocks.

The triangular diagram shows the Fe–Ti–O minerals (Fig. 3.9). Although magnetite, ilmenite and haematite are usually considered to be the common examples, precise identification may be difficult due to extensive chemical substitution within the Fe–Ti–O system, as well as the presence of Cr, Mn, Mg and Al in these minerals. The Fe–Ti–O minerals often occur in frequently submicroscopic intergrowths, which result from cooling and oxidation/reduction.

In typical basaltic igneous rocks there are two primary oxide minerals, ferrianilmenite and titanomagnetite. On slow cooling ferrianilmenite may become ilmenite with haematite lamellae, whereas titanomagnetite may produce lamellae of ilmenite before breaking down to a fine intergrowth of ulvospinel and magnetite.

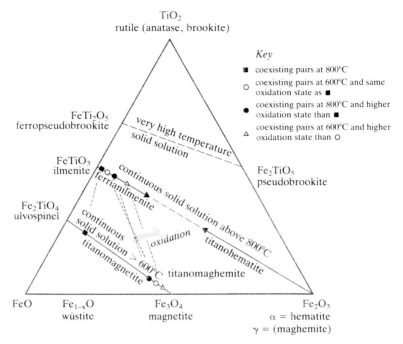

Figure 3.9   The iron–titanium oxide minerals.

Regionally metamorphosed sediments typically contain assemblages of almost pure magnetite ± ferrianilmenite (possibly with exsolved haematite) ± titanohaematite (possibly with exsolved ilmenite) ± rutile. Greenschist facies rocks may contain a magnetite + rutile assemblage which gives way to a titanohaematite + ferrianilmenite assemblage in amphibolite facies.

The following are noted for Figure 3.9:

(a) TiO_2 polymorphs are:

rutile, tetragonal $c/a < 1$
anatase, tetragonal $c/a > 1$ (metastable?) ⎫ found in low-
⎪ temperature
brookite, orthorhombic   (metastable?) ⎪ hydrothermal
⎭ environment

(b)   Ferropseudobrookite is only stable above 1100 °C and is very rare.

(c)   Pseudobrookite is only stable above 585 °C and is rare (in high-temperature contact metamorphosed rocks).

(d)   The ilmenite–haematite solid solution series is complete above about 800 °C.

(e)   The ulvospinel–magnetite solid solution series is complete above about 600°.

(f) Magnetite and rutile can coexist only below about 400°C.

(g) From 1100°C to 600°C, Ti-rich ferrianilmenite coexists with Ti-poor titanomagnetite; the exact composition of the coexisting pair depends on the oxygen fugacity as well as the temperature (the Buddington & Lindsley (1964) magnetite–ilmenite geothermometer oxygen barometer). The dotted lines show how the (approximate) compositions of coexisting pairs depend on temperature and oxygen fugacity.

(h) Oxidation of titanomagnetite at relatively high temperatures results in exsolution lamellae of ferrianilmenite in the (111) orientation in magnetite, and this oxidation can result from cooling alone. Similarly, reduction of ferrianilmenite results in titanomagnetite lamellae in the (0001) orientation of ilmenite.

(i) Titanomaghemites form at low temperatures (< 600°C) by non-equilibrium oxidation of titanomagnetites; they are cation deficient and have a wide range of compositions.

(j) Hemo-ilmenite is a ferrianilmenite host with titanohaematite lamellae.

(k) Ilmenohaematite is a titanohaematite host with ferrianilmenite lamellae.

(l) The *bulk* composition of coexisting hemo-ilmenite and ilmeno-haematite grains depends on temperature.

(m) Wüstite is cation deficient relative to FeO. It is very rare, as it is stable only above 570°C at low oxygen fugacities.

Spinel group

The general unit cell formula of the spinels is $R_8^{2+}R_{16}^{3+}O_{32}$, where R^{2+} and R^{3+} represent divalent and trivalent cations respectively, but the formula is usually simplified to R_3O_4. All spinels are cubic, but there are two structural types with differing distributions of the cations:

Normal spinels

R_8^{2+} in fourfold tetrahedral coordination with oxygen
R_{16}^{3+} in sixfold octahedral coordination with oxygen

Inverse spinels

R_8^{3+} in fourfold tetrahedral coordination with oxygen
R_8^{2+} and R_8^{3+} in sixfold octahedral coordination with oxygen

Most natural spinels have an intermediate structure (see Fig. 3.10). In the spinel structure, oxygen is O^{2-} in tetrahedral coordination.

The spinels are normal valence compounds in that the total cation charge balances the total anion charge. Divalent R^{2+} cations include Mg^{2+}, Fe^{2+}, Zn^{2+}, Mn^{2+} and Ni^{2+}; and R^{3+} cations include Al^{3+}, Fe^{3+} and Cr^{3+}. One way of representing the extensive solid solution in spinels is shown in Figure 3.11.

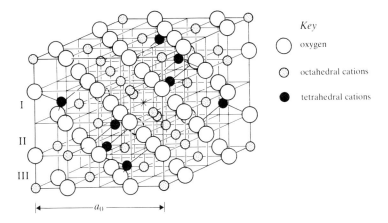

Figure 3.10 The spinel unit cell, orientated so as to emphasize the (111) planes. Atoms are not drawn to scale; the circles simple represent the centres of atoms (after Lindsley, in Rumble 1976).

It is possible for Ti^{4+} (and V^{4+}) to enter the structure due to a coupled substitution of the type $2Fe^{3+} \rightleftharpoons Fe^{2+} + Ti^{4+}$. The inverse spinel structure of maghemite γ-Fe_2O_3 supports a cation site vacancy which is produced by the substitution $3Fe^{2+} \rightleftharpoons 2Fe^{3+} + [\ ]$; the formula may be written:

$$Fe_3^{3+}(tetr.)Fe_3^{3+}(oct.)Fe_2^{3+}[\ ](oct.)O_{12}$$

Transparent spinels have high relief ($n > 1.7$) in thin section and are isotropic. An octahedral habit, sometimes with twinning on {111}, helps to distinguish them from the garnets. Opaque spinels are isotropic; they differ in their reflectance values.

There is considerable variation in the chemical composition of natural spinels, and this leads to a variety of colours and degrees of

Figure 3.11
Solid solution
in spinels.

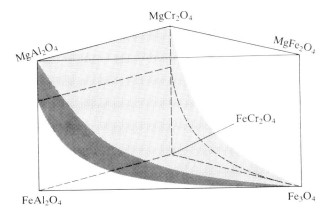

Table 3.2 Spinels.

Spinel	Composition	Structural type	Colour/opacity
spinel	$MgAl_2O_4$	normal	$n = 1.719$, colourless or red, blue brown etc.
gahnite	$ZnAl_2O_4$	normal	$n \approx 1.8$, dark green to brown
hercynite	$FeAl_2O_4$	normal	$n = 1.835$, dark green to black
magnetite	Fe_3O_4	inverse	opaque
maghemite	$\gamma\text{-}Fe_2O_3$	inverse	opaque (metastable with respect to haematite)
ulvospinel	Fe_2TiO_4	inverse	opaque
chromite	$FeCr_2O_4$	normal	opaque, dark brown on edges
Common intermediate varieties			
pleonaste	$(Mg,Fe)Al_2O_4$	normal	green to blue green
picotite	$(Fe,Mg)(Al,Cr)_2O_4$		

opacity (Table 3.2). Magnetic (ferrimagnetic) spinels also exist, the best known being magnetite. The spinels can often be identified on the basis of their mode of occurrence, textural relations and associated phases, but chemical analysis is usually required for satisfactory identification.

Cassiterite SnO_2

Cassiterite may contain minor amounts of Fe, Nb, Ta, Ti, W or Si.

Crystals Cassiterite is tetragonal, with $a:c = 1:0.672$. Crystals are usually short prisms [001], with {110} and {100} prominent (Fig. 3.12). Faces in [10$\bar{1}$] and [001] are often striated. Twinning on {011} is very common, giving both contact and penetration twins. There is a poor {100} cleavage.

cassiterite twin
on (011)

Figure 3.12 Cassiterite crystals.

Thin section	$n_o = 1.990–2.010$
	$n_e = 2.093–2.100$
	$\delta = 0.096–0.098$
	Uniaxial + ve (crystals are length slow)
	$D = 6.98–7.02 \qquad H = 6\frac{1}{2}$

COLOUR — Colourless or slightly red or brown.

*PLEOCHROISM — Occasionally present in coloured varieties, with o pale colours and e dark yellow, brown or reddish.

CLEAVAGE — {100} and {010} prismatic cleavages parallel to the length of the mineral.

*BIREFRINGENCE — Very high, but interference colours are often masked by the colour of the mineral.

TWINNING — Simple and repeated, common on {011}.

ZONING — Variation in the iron content leads to colour banding.

Polished section — Cassiterite is grey, sometimes appearing slightly brownish. With $R_o = 11\%$ and $R_e = 12\%$, bireflectance is weak, but usually visible in granular aggregates and twinned grains. Cassiterite is darker than sphalerite and only slightly brighter than gangue minerals. Anisotropy is distinct in greys, but internal reflections, which are common and colourless to brown, often mask the anisotropy.

Cassiterite occurs as isolated prismatic to rounded crystals, geniculate (knee-like) twins or granular aggregates. Colloform aggregates containing colloidal haematite are known as "wood tin". Twinning is common, and cleavage traces are often present. Zonation of the iron content may be seen in crossed polars, because Fe absorbs the light in the internal reflections. VHN = 1240–1470.

Occurrence — Cassiterite is mainly found with wolframite, tourmaline, topaz, arsenopyrite, molybdenite, pyrrhotite and bismuthian minerals in high-temperature hydrothermal veins, pegmatites, greisens, stockworks and disseminations associated with acid igneous rocks. It is found as a detrital heavy mineral in sediments (such as the commercial placer deposits of Malaysia) and in gossans over stanniferous sulphide deposits. Wood tin is found in the secondary oxidation zone.

Distinguishing features — Compared with cassiterite, sphalerite is brighter, isotropic and softer; wolframite is slightly brighter and has fewer internal reflections; and rutile is brighter.

Chromite — $FeCr_2O_4$

Crystals — Usually containing Mg and Al, chromite may also contain Zn, V and Mn.

Thin section — Chromite is cubic, and is a member of the spinel group. Crystals are rare, but occur as octahedra modified by {001} faces. There is no cleavage. $D = 5.1$, $H = 6\frac{1}{2}$.

Chromite is opaque, except in very thin grain margins which are brownish in colour.

Polished section Chromite is grey, sometimes appearing slightly brownish. $R = 13\%$, but varies with chemical composition. This reflectance value is significantly less than that of magnetite. High Fe and Cr values increase R, but Al and Mg decrease R. Although cubic and usually isotropic, chromite sometimes shows weak anisotropy. Iron-poor chromite may have scarce reddish-brown internal reflections.

Chromite occurs as rounded octahedral grains resembling drop-lets, interstitially in silicates, or as granular aggregates. It is an accessory mineral in most peridotites and derived serpentinites. Cataclastic texture is common. A zonation in reflectance, related to chemical zonation, may be observed. Marginal discoloration and alteration may occur. Inclusions of Fe + Ti + O phases, e.g. rutile, may be present. VHN = 1270–1460.

Occurrence Chromite is only abundant in certain mafic igneous rocks, especially large layered intrusions (e.g. the Bushveldt lopolith) as cumulates or possibly oxide–liquid segregations. It is found as podiform con-centrations, possibly originally cumulates, in Alpine-type serpenti-nites, and also as a detrital heavy mineral in sedimentary and metamorphic rocks. Chromite may occur as cores within magnetite grains. Iron-rich rims of chromites, commonly observed in serpenti-nites, are known as ferrit-chromit. The rims have a slightly higher reflectance than the chromite cores and are magnetic.

Distinguishing features Compared with chromite, magnetite is brighter. The two minerals are similar unless direct comparison of brightness can be made. However, remember that magnetite is magnetic!

Corundum Al_2O_3 trigonal, c/a 1.364
$n_o = 1.768–1.772$
$n_e = 1.760–1.763$
$\delta = 0.008–0.009$
Uniaxial − ve (crystal hexagonal, rarely prismatic)
$D = 3.98–4.02$ $*H = 9$

COLOUR Colourless, but gem-quality corundum is often coloured blue (sapphire) or red (ruby) in hand specimen.

PLEOCHROISM Normal corundum is not pleochroic, but gem-quality minerals are weakly pleochroic, particularly sapphire with e blue and o light blue.

HABIT Rarely euhedral, usually as small rounded crystals.
*CLEAVAGE None; basal parting present.
*RELIEF High (about the same as garnet).
ALTERATION Corundum can alter to Al_2SiO_5 minerals during metamorphism, by addition of silica, or to muscovite if water and potassium are also available.

Figure 3.13
Typical
haematite
crystals.

BIREFRINGENCE	Low, common interference colours are first-order greys and whites.
*TWINNING	Lamellar twinning is commonly seen on $\{10\bar{1}1\}$. Simple twins can occur, with $\{0001\}$ as the twin plane.
DISTINGUISHING FEATURES	Corundum has low birefringence, high relief, no cleavage and lamellar twinning. Apatite has lower relief and still lower birefringence.
*OCCURRENCE	Corundum occurs in silica-poor rocks such as nepheline–syenites, and other alkali igneous undersaturated rocks. It may occur in contact aureoles in thermally altered aluminous shales, and in aluminous xenoliths found within high-temperature basic igneous plutonic and hypabyssal rocks. In these aluminous xenoliths, corundum is frequently found in association with spinel, orthopyroxene and cordierite. Corundum occurs in metamorphosed bauxite deposits, and also in emery deposits along with **högbomite** $Mg(Fe^{3+},Ti,Al)_4O_7$, spinel and magnetite. It can occur as a detrital mineral in sediments.

Haematite	Fe_2O_3
	Haematite is often titaniferous, i.e. there is a haematite–ilmenite solid solution (see Section 3.4).
Crystals	Haematite is hexagonal, $a:c = 1:1.3652$ and it usually occurs as tabular crystals $\{0001\}$, often in subparallel growths (Fig. 3.13). Penetration twinning occurs on $\{0001\}$ and lamellar twinning on $\{10\bar{1}1\}$. There is no cleavage. $D = 5.2$, $H = 6\frac{1}{2}$.
Thin section	Haematite is opaque, but deep red in very thin plates. It is uniaxial negative, with absorption $o > e$.
Polished section	Haematite is light grey and only weakly bireflecting, with $R_o = 30\%$ and $R_e = 26\%$. It is much brighter than magnetite and ilmenite. Anisotropy is strong in bluish and brownish greys. The deep-red internal reflections are scarce except in very thin plates. Haematite coatings give a red coloration to internal reflections of transparent grains such as quartz.

Haematite occurs as idiomorphic tabular crystals and fibrous radiating aggregates (Fig. 3.14). It is also found as microcrystalline colloform masses. It is often intergrown with other Fe + Ti + O minerals and it occurs as lamellae in ilmenite. Haematite may contain lamellae of ilmenite or rutile. Lamellar twinning is common, and a pseudo-cleavage consisting of elongate pits may be present. VHN = 1000–1100.

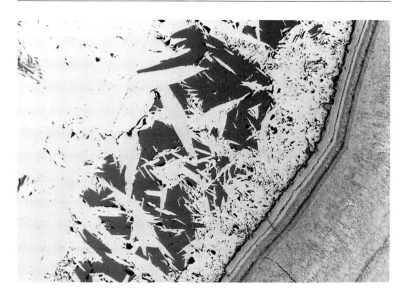

Figure 3.14 Haematite plates (light grey) in quartz (dark grey). Colloform haematite/quartz is on the right. Note the lack of distinct bireflectance of the hematite. Compare with stibnite (R-PPL).

Occurrence Haematite is found with other Fe–Ti–O minerals in igneous and metamorphic rocks as well as sedimentary rocks, especially banded iron formations. Haematite in veins can be primary, but it frequently forms by oxidation of other primary iron-bearing minerals, in gossans for example.

Distinguishing features Compared with haematite, stibnite has a distinct bireflectance, is softer and has a good cleavage; ilmenite is pinkish and darker; and cinnabar has abundant internal reflections and is softer.

Note Martite is magnetite pseudomorphed by an intergrowth of haematite (Fig. 3.15).

Ilmenite $FeTiO_3$
Ilmenite may contain Mn or Mg, the magnesian end member being geikielite and the manganiferous end member being pyrophanite. It may also contain Fe^{3+}, which represents a solid solution towards haematite Fe_2O_3.

Crystals Ilmenite is trigonal, $a:c = 1:1.3846$, and occurs as tabular $\{0001\}$ crystals. Twinning occurs on $\{0001\}$, and multiple twinning on $\{10\bar{1}1\}$. There is no cleavage, but there is a parting parallel to $\{10\bar{1}1\}$. $D = 4.7$, $H = 5\frac{1}{2}$.

Thin section In very thin flakes ilmenite is red, and unixial negative.

Polished section Ilmenite is slightly pinkish or brownish light grey, with a weak pleochroism. $R_o = 20\%$, which is similar to magnetite, and

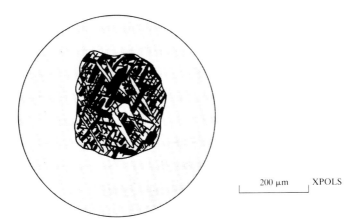

200 μm   XPOLS

Figure 3.15 Martite: a "heavy mineral" grain of magnetite replaced by haematite.

R_e = 17%. Anisotropy is moderate, but is only distinct in some orientations; tints are greenish, bluish and brownish greys.

Ilmenite is sometimes idiomorphic, but is usually intergrown with other Fe–Ti–O minerals. It often contains lamellar inclusions of haematite (Fig. 3.16) or other Fe–Ti–O minerals. Occasionally, lamellar twins may be present. VHN = 560–700, varying with chemical composition.

Occurrence Ilmenite is found with other Fe–Ti–O minerals in igneous rocks (especially of mafic composition) and metamorphic rocks, and also (but rarely) in veins and pegmatites. Detrital ilmenite is usually altered to leucoxene, which is enriched in TiO_2. It occurs in heavy-mineral concentrates. Magnesium-rich ilmenites occur in kimberlites, but also in contact metamorphosed rocks.

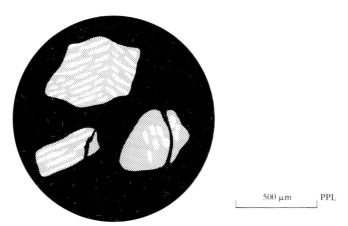

500 μm   PPL

Figure 3.16 Grains of ilmenite with exsolved haematite.

181

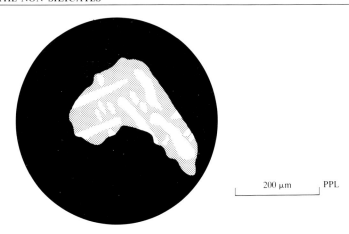

200 μm   PPL

Figure 3.17   A grain of magnetite with exsolved ilmenite

Distinguishing features   Compared with ilmenite, magnetite is slightly brighter and usually bluish-grey in direct comparison, isotropic and strongly magnetic. Rutile shows abundant internal reflections.

Magnetite Fe_3O_4
Magnetite often contains Ti, Cr or Mn. Titaniferous magnetite often contains ulvospinel Fe_2TiO_4 in solid solution.

Crystals   Magnetite is an inverse spinel. It is cubic, commonly occurring as octahedra and combinations of the octahedron and rhombic dodecahedron. Twinning is common on {111}, the usual spinel twin. $D = 5.2$, $H = 5\frac{1}{2}$

Polished section   Magnetite is grey, sometimes with a brownish or pinkish tint indicative of titanium (ulvospinel is brownish grey). $R = 21\%$, making magnetite much darker than pyrite and haematite. Magnetite is isotropic, with good extinction.

It is often found as idiomorphic octahedral sections, but also as skeletal grains or granular aggregates. Lamellae of haematite are often in a triangular pattern. Lamellae (Fig. 3.17) and blebs of ilmenite in a fine "frosty" texture of ulvospinel in magnetite represent slowly cooled titaniferous magnetite. Also, exsolved lamellae and blebs of dark grey spinels may be present. VHN = 500–790.

Occurrence   Magnetite is usually found with other Fe–Ti–O minerals in igneous and metamorphic rocks and skarns. It also occurs as a heavy mineral in sediments and sedimentary rocks, and in high-temperature hydrothermal veins with sulphides. It represents reducing conditions relative to haematite.

Distinguishing features   Compared with magnetite, ilmenite is similar but often pinker and anisotropic; sphalerite is softer, usually has internal reflections and occurs in a different association; and chromite is very similar in

182

Figure 3.18
A rutile crystal
and twinning.

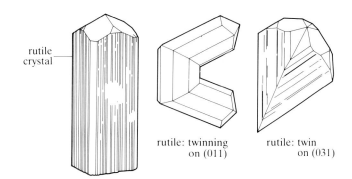

rutile
crystal

rutile: twinning
on (011)

rutile: twin
on (031)

isolation but is darker and may show internal reflections. A magne-
tized needle may be used to confirm the magnetism of magnetite
grains in polished section.

Rutile TiO_2
Rutile may contain some Fe or Nb. The polymorphs anatase and
brookite are almost identical to rutile in polished section.

Crystals Rutile is tetragonal, $a:c = 1:0.6442$. Crystals are commonly prismatic, often slender to acicular (Fig. 3.18). Twinning on {011} is
common and is often repeated or geniculated. There is a distinct
cleavage on {110}, as well as a cleavage on {100}.

Thin section $n_o = 2.605–2.613$
$n_e = 2.899–2.901$
$\delta = 0.286–0.296$
Uniaxial + ve (crystals are length-slow)
$D = 4.23–5.5$ $H = 6–6\frac{1}{2}$

*COLOUR Reddish-brown or yellowish (sometimes opaque).
PLEOCHROISM Common, although sometimes weak, with o pale yellow, and e pale
brown, dark brown or red.
HABIT Small, acicular prisms are common.
CLEAVAGE {110} and {100} prismatic cleavages are good.
*RELIEF Exceptionally high.
*BIREFRINGENCE Extremely high; interference colours usually masked by mineral
colour.
TWINNING Common in a number of different planes.
Polished Rutile is light grey with a slight bluish tint. $R_o = 20\%$ and
section $R_e = 23\%$. The bireflectance is weak but usually distinct. Rutile has
about the same brightness as magnetite. It is strongly anisotropic in
greys, but the anisotropy is often masked by abundant bright
colourless yellow to brown internal reflections. In iron-rich varieties, internal reflections are less abundant and reddish.
Rutile occurs as prismatic to acicular isolated crystals or as

183

Figure 3.19
A spinel crystal
and twinning.

spinel crystal   spinel twin on
(111)

aggregates of crystals and in spongy porphyroblasts. It usually occurs as small grains. Multiple and simple twins are common. VHN = 890–970.

Occurrence   Rutile is associated with other Fe–Ti–O phases in pegmatites, igneous and metamorphic rocks. It is a heavy mineral in sediments. It is often produced from ilmenite on wall-rock alteration by hydrothermal solutions (e.g. greisenization). Rutile occurs within quartz crystals as long thread-like crystals.

Distinguishing features   Compared with rutile, haematite is whiter and brighter and rarely shows internal reflections; ilmenite is slightly pinkish and does not show internal reflections; and cassiterite tends to be more equant and is darker.

Notes   The low-temperature TiO_2 polymorphs anatase (uniaxial negative) and brookite (orthorhombic, small positive $2V$) have a similar occurrence to rutile. In polished section, anatase lacks twinning and is only very weakly anisotropic. Anatase occurs associated with clay minerals in sedimentary rocks.

Spinel $MgAl_2O_4$

The two common aluminous spinels are spinel $MgAl_2O_4$ and hercynite $FeAl_2O_4$. The general formula of aluminous spinels is $M^{2+}M_2^{3+}O_4$, with M^{2+} = Mg, Fe, Mn, Zn and M^{3+} = Al. There is extensive solid solution including $Al^{3+} \rightleftharpoons Fe^{3+}$, Cr^{3+}.

Crystals   Spinel is cubic and usually of octahedral habit (Fig. 3.19). Twinning on {111} may be repeated. There is a poor {111} cleavage. $D = 3.55$, $H = 6$–8 ($D = 4.40$, $H = 7\frac{1}{2}$ for hercynite).

Thin section   Spinel is of variable colour and opacity. Mg-rich spinel is transparent and isotropic.

Polished section   Spinel is grey, with $R = 8\%$, making it only slightly brighter than associated silicates. It is isotropic. Internal reflections vary in abundance depending on the composition.

Spinel is often idiomorphic or rounded octahedral. It may contain inclusions of magnetite or ilmenite. VHN = 861–1650 (spinel), 1402–1561 (hercynite).

Occurrence   Spinel occurs as exsolved blebs or lamellae in magnetite. It is found in basic igneous rocks, and in contact metamorphic and metaso-

184

500 μm   PPL

Figure 3.20 Pitchblende: note the "patchiness' of the brightness due to variation in oxidation. Shrinkage cracks radiate from the centre of spheroids.

matic aluminous (or Si-deficient) rocks. It is also found as a heavy mineral in placer deposits. Unlike spinel, hercynite is stable in the presence of free silica.

Uraninite UO_2

Natural uraninite is often oxidized to some extent to pitchblende UO_{2-3}. The U is often replaced by Th or Ce.

Crystals Uraninite is cubic and usually occurs as octahedra, cubes or dodecahedra. Twinning on {111} is rare. There is no cleavage. $D \approx 9.0$, $H = 6\frac{1}{2}$.

Thin section Uranium oxides often appear as opaque rounded aggregates, altered along fractures. In thin splinters a green to brown colour may be obtained. Associated minerals may be darkened due to radiation damage.

Polished section Uraninite is grey, with $R = 17\%$, similar to sphalerite. It is cubic and isotropic. Pitchblende is similar but slightly darker, with $R = 16\%$. Scarce brown internal reflections may be observed in these minerals.

Uranium oxides commonly occur as spherical or botryoidal masses. Uraninite is well crystallized, but pitchblende varies in crystallinity and non-stoichiometry and tends to polish poorly. Composition zoning results in slight brightness and hardness changes. Shrinkage cracks occur in pitchblende (Fig. 3.20). VHN = 782–839 (uraninite), 500–550 (pitchblende).

Occurrence Uranium oxides are found in high-temperature pegmatitic to low-temperature hydrothermal vein and replacement deposits. There is an association with Ni + Co + Ag + Bi mineralization, with acid

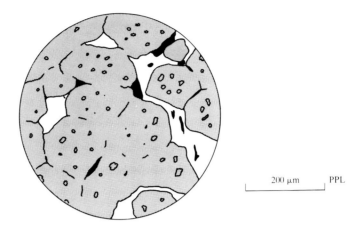

Figure 3.21 Thucolite: carbon (dark grey) with inclusions of uraninite (dark grey). Pyrite and gold (both white) are interstitial. This is typical of cross sections of Witwatersrand columnar thucolite.

igneous rocks and with organic material in sedimentary rocks. Detrital uraninite is found in placer deposits with gold.

Distinguishing Compared with uraninite, magnetite is similar but is magnetic. The
features uranium oxides are radioactive!
Notes Oxidation of primary uranium oxides often results in distinctive bright yellow or green secondary uraniferous minerals.

Thucolite is fragmental uraninite in polymerized carbonaceous material (Fig. 3.21).

3.8 Phosphates

Apatite $Ca_5(PO_4)_3(OH,F,Cl)$ hexagonal, c/a 0.735
$n_o = 1.623–1.667$
$n_e = 1.624–1.666$
$\delta = 0.001–0.007$
Uniaxial $-$ ve (a prism section is length fast)
$D = 3.1–3.35$ $H = 5$

COLOUR Colourless.
*HABIT Small prismatic crystals with hexagonal cross section, often found with ferromagnesian minerals in rocks, particularly amphiboles and micas.
CLEAVAGE Good basal {0001} cleavage; imperfect prismatic cleavage {1010}.
RELIEF Moderate.

186

*BIREFRINGENCE	Very low; maximum interference colours are grey.
OCCURRENCE	Important accessory mineral in igneous rocks, especially acidic plutonic rocks, granite pegmatites and vein rocks, but also common in diorites and gabbros.

Apatite is common in metamorphic rocks, especially chlorite schists and amphibole-bearing schists and gneisses.

Apatite occurs as a detrital mineral in sedimentary rocks. Sedimentary phosphatic deposits commonly contain a cryptocrystalline phosphatic mineral called "collophane", a term used if apatite cannot be positively identified.

Monazite (Ce,La,Th)PO$_4$

monoclinic
$0.969:1:0.922, \beta = 103°38'$

$n_\alpha = 1.770–1.800$
$n_\beta = 1.777–1.801$
$n_\gamma = 1.828–1.851$
$\delta = 0.045–0.075$
$2V_\gamma = 6°–19° + ve$
OAP is approximately perpendicular to the a axis
$D = 4.6–5.4 \qquad H = 5–5\frac{1}{2}$

*COLOUR	Colourless, often pale yellow.
*PLEOCHROISM	Weakly pleochroic in yellows with α and γ pale yellows, and β yellow.
HABIT	Blocky, euhedral crystals common, often elongated along the b axis. Seen as detrital grains in sediments.
CLEAVAGE	Distinct on {010}; other cleavages poor, but a distinct basal {001} parting is usually present.
*ALTERATION	None.
*RELIEF	Very high.
BIREFRINGENCE	Bright colours of high orders seen in sections perpendicular to the a axis (roughly (100) sections). Basal sections show weak birefringence.
*INTERFERENCE FIGURE	A basal section shows a very small positive Bx$_a$ figure.
EXTINCTION ANGLE	An (010) section shows slow $(\gamma)^\wedge$cleavage = 2°–70°.
*OCCURRENCE	Monazite occurs as tiny euhedral accessory minerals in granites, and other similar igneous rocks. Large crystals may be present in late-stage pegmatites with xenotime, apatite, zircon, columbite etc.; and Alpine-type veins may contain monazite in addition to anatase, wolframite and cassiterite. It is rare in metamorphic schists and gneisses, but is a common mineral in detrital sands derived from some of the rock types mentioned above.

187

Xenotime YPO_4 tetragonal, c/a 0.875
$n_o = 1.719–1.724$
$n_e = 1.816–1.827$
$\delta = 0.095–0.107$
Uniaxial $+$ ve (the crystal is length slow)

COLOUR	Colourless to pale yellow green or yellowish brown.
*PLEOCHROISM	Weakly pleochroic, with o pale pink, pale yellowish-brown or pale yellow, and e yellow, brown or yellowish-green.
HABIT	Euhedral crystals common, elongated on the c axis; it has a shape similar to zircon.
CLEAVAGE	Distinct {110} prismatic cleavages intersect at right angles in a basal section.
RELIEF	High to very high.
*ALTERATION	None.
*BIREFRINGENCE	Very high to extreme, with fifth and higher orders seen.
OCCURRENCE	Xenotime is a widespread accessory mineral in granites, and syenites and other similar igneous rocks. *It is frequently mistaken for zircon.* Large crystals may occur in some complex granite pegmatites with zircon, monazite, apatite, allanite and other minerals containing REES. It may occur in gneisses and is a common heavy mineral in detrital sediments and placer deposits.

Wavellite $Al_3(OH)_3(PO_4)_2.5H_2O$ orthorhombic
0.5577 : 1 : 0.4061

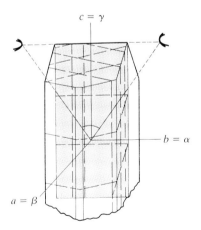

$n_\alpha = 1.518–1.535$
$n_\beta = 1.526–1.543$
$n_\gamma = 1.545–1.561$
$\delta = 0.025–0.027$

$2V_\gamma \approx 72°$ + ve

OAP is parallel to (100) (a prism section is length slow)

$D = 2.32–2.37 \qquad H = 3\frac{1}{2}–4$

COLOUR — Usually colourless, but may be coloured in blue, green, yellow and brown.

PLEOCHROISM — When present pleochroism may be marked, with α deep blue or deep green, β yellowish-brown, and γ colourless, pale yellow or pale brown.

*HABIT — Single acicular crystals are rare, and wavellite usually occurs as spherical masses in veins or cavities.

*CLEAVAGE — Prismatic cleavages are perfect on {110}, good on {101}, and distinct on {010}.

RELIEF — Low.

ALTERATION — None.

BIREFRINGENCE — Maximum low second-order colours are seen in (100) section.

INTERFERENCE FIGURE — Basal sections give Bx_a figures with a large $2V$, but a single optic axis is required for the sign.

*OCCURRENCE — Wavellite is a widespread secondary mineral deposited by hydrothermal solutions and found in fractures in limonite ores, phosphatic rocks and low-grade aluminous metamorphic rocks. It may occur in pegmatites and high-temperature veins with cassiterite and apatite.

3.9 Sulphates

Anhydrite CaSO$_4$ orthorhombic
0.999 : 1 : 0.892

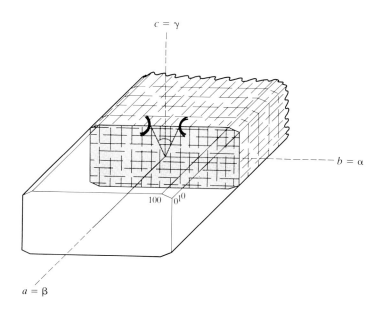

$n_\alpha = 1.569\text{–}1.574$
$n_\beta = 1.574\text{–}1.579$
$n_\gamma = 1.609\text{–}1.618$
$\delta = 0.04$
$2V_\gamma = 42°\text{–}44°\ +\text{ve}$
OAP is parallel to (010) (the crystal is elongated along the a axis and can be length fast or slow)
$D = 2.9\text{–}3.0 \qquad H = 3\text{–}3\frac{1}{2}$

COLOUR	Colourless
HABIT	Prismatic crystals, with aggregates common.
CLEAVAGE	{010} perfect, {100} and {001} good.
RELIEF	Low to moderate.
ALTERATION	Just as gypsum can dehydrate to anhydrite, so anhydrite can react with water to form gypsum.
*BIREFRINGENCE	High (much greater than gypsum) with third-order colours.
*INTERFERENCE FIGURE	Seen on an elongate prism section, a Bx$_a$ figure is just larger than the field of view and positive.
*EXTINCTION	Straight on all cleavages and prism edges.
TWINNING	Repeated on {011}.

*OCCURRENCE   Similar to gypsum, being found in evaporite deposits. Anhydrite may form by hydrothermal alteration of limestones or dolomites.

Barite (baryte or barytes)
> $BaSO_4$ orthorhombic

Celestite (celestine)
> $SrSO_4$ 1.629:1:1.312 (ba)
>   1.562:1:1.283 (ce)

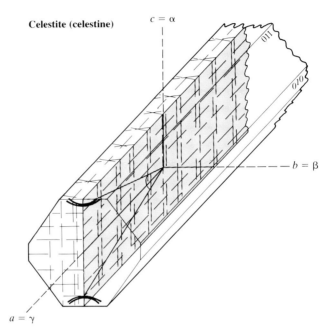

Celestite (celestine)

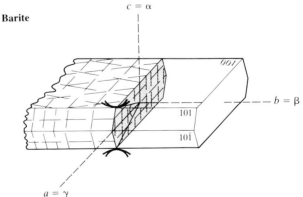

Barite

n_α	1.634 (Ba)	1.621 (Cel)
n_β	1.636	1.623

191

n_γ 1.646   1.630
δ 0.012   0.009
$2V_\gamma$ 37° + ve $\approx 50°$ + ve (the crystal is length slow)
OAP is parallel to (010): celestite is elongated along the a axis
$D = 4.5$ $H = 2\frac{1}{2}-3\frac{1}{2}$ (barite)
$D = 4.0$ $H = 2\frac{1}{2}-3\frac{1}{2}$ (celestite)

COLOUR   Colourless.
HABIT   Subhedral clusters of prismatic crystals common
CLEAVAGE   Basal cleavage {001} perfect; {210} and {010} cleavages present.
RELIEF   Moderate.
BIREFRINGENCE Low, first-order yellows, but mottled colours are common.
*INTERFERENCE Bx_a figure with small $2V$ seen on (100) section, i.e. section with two
FIGURE   cleavages.
EXTINCTION   Straight on cleavage trace or prism edge.
TWINNING   Lamellar twinning present on {110}.
*OCCURRENCE   Barite occurs as a gangue mineral in ore-bearing hydrothermal veins, in association with fluorite, calcite and quartz. Barite occurs as stratiform deposits of synsedimentary exhalative origin in sedimentary and metamorphic terrains.

Celestite $SrSO_4$ is similar to barite optically and occurs in dolomites, in evaporite deposits, and rarely in hydrothermal veins. There is probably a complete solid solution series from barite to celestite.

Gypsum $CaSO_4.2H_2O$ monoclinic

$0.374:1:0.414, \beta = 113°50'$

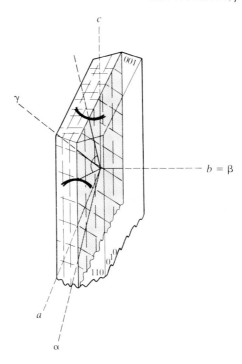

$n_\alpha = 1.519–1.521$
$n_\beta = 1.523–1.526$
$n_\gamma = 1.529–1.531$
$\delta = 0.01$
$2V_\gamma \approx 58° + ve$
OAP is parallel to (010)
$D = 2.30–2.37$ $H = 2$

COLOUR   Colourless.
HABIT   Anhedral crystals occur usually in aggregate masses.
CLEAVAGE   {010} perfect, {100} and {011} good.
RELIEF   Low, always less than CB.
ALTERATION With increase in temperature gypsum changes to anhydrite thus
            (about 200°C):

$$CaSO_4.2H_2O \rightarrow CaSO_4 + 2H_2O$$

BIREFRINGENCE Low, interference colours are first order whites.
INTERFERENCE   A Bx_a figure is seen on a thin prismatic section, but $2V$ is larger than
FIGURE   the field of view; thus the sign is best determined by looking at an
               optic axis figure.
EXTINCTION   Straight on the {010} cleavage.

193

Figure 3.22

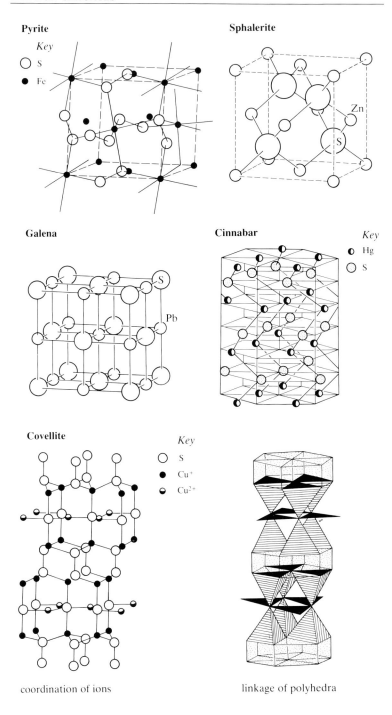

Pyrite

Key
○ S
● Fe

Sphalerite

Zn
S

Galena

S
Pb

Cinnabar

Key
◑ Hg
○ S

Covellite

Key
○ S
● Cu⁺
◑ Cu²⁺

coordination of ions

linkage of polyhedra

Tetrahedrite $Cu_{10}Zn_2Sb_4S_{13}$ $^1/_2$ cell
(after Pauling & Neuman 1934)

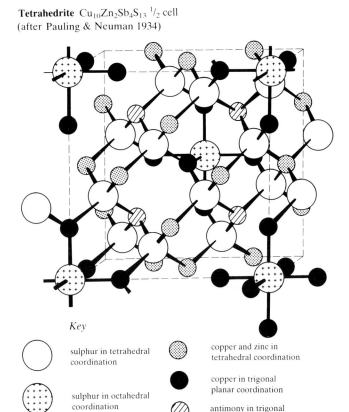

Key

◯ sulphur in tetrahedral coordination

⬚ sulphur in octahedral coordination

⬤ copper and zinc in tetrahedral coordination

⬤ copper in trigonal planar coordination

⬱ antimony in trigonal pyramidal coordination

Figure 3.22
Sulphide
structures
(after Vaughan
& Craig 1978).

TWINNING   Common on {100}; repeated twinning usually seen.

*OCCURRENCE   Gypsum is mainly found in sedimentary rocks, especially in evaporitic sequences.

Calcium sulphate can occur as either gypsum or anhydrite. Anhydrite may be formed by the dehydration of primary gypsum. In desert regions calcium sulphate is dissolved in percolating ground waters, which can be drawn to the surface and deposit gypsum as "desert roses" during very dry spells.

Gypsum can form in fissures in shales and other argillaceous rocks by the action of acid groundwaters (sulphuric acid in solution), reacting with calcium either from limestone nodules within the argillaceous rocks or from intercalated limestone beds.

3.10 Sulphides

In the structures of sulphide minerals, sulphur atoms are usually surrounded by metallic atoms (e.g. Cu, Zn or Fe) or the semi-metals (Sb, As or Bi). The chemical bonding is usually considered to be essentially covalent. Although sulphur has a preference for fourfold tetrahedral coordination, it is found in a large variety of coordination polyhedra which may be quite asymmetric. Non-stoichiometry, i.e. a variable metal:sulphur ratio, is a feature of many sulphide structures, especially at high temperatures; complex ordering may result on cooling of a non-stoichiometric phase leading, at low temperature, to minerals with only slightly different compositions but different structures. A good example is that of high-temperature cubic digenite, $Cu_{2-x}S (x \leqslant 0.2)$, which is represented at low temperatures by orthorhombic chalcocite Cu_2S, orthorhombic djurleite $Cu_{1.97}S$ and cubic digenite $Cu_{1.8}S$.

Two further possible complexities in sulphide structures are the existence of sulphur–sulphur bonds, exemplified by the S_2^{2-} pair in pyrite FeS_2 (see Fig. 3.22, which emphasizes the octahedral site of iron), and the existence of structures that can be considered as resulting from a replacement by a semi-metal of half the sulphur in such pairs, e.g. arsenopyrite FeAsS.

Most sulphides are opaque but some (e.g. sphalerite when pure zinc sulphide) are transparent. Some are transparent for red light (e.g. pyrargyrite Ag_3SbS_3) or only in the infrared (e.g. stibnite Sb_2S_3). Many are semiconductors, which means that they conduct electricity at a high temperature but not at a low temperature. In fact, the optical and physical properties of many sulphides are best understood if the band model of semiconductors is applied (see Shuey 1975).

The structures of several common sulphides are illustrated in Figure 3.22. As is evident from the few examples given, sulphide structures can be classified – as are the silicates – into structures based on chains, sheets, networks and so on. Although such a classification is of less value than for the silicates, consideration of structures in such a way helps to explain the crystal morphology, cleavage directions etc. of some sulphides.

The sulphosalts are one group of sulphides which are very diverse chemically and structurally. They contain a semi-metal as well as a metal and sulphur in their structures; the semi-metal is typically bonded to sulphur in trigonal pyramidal coordination, but there is no semi-metal to metal bond as in arsenopyrite FeAsS. Two examples of sulphosalts which are relatively common are pyrargyrite Ag_3SbS_3 and tetrahedrite $(Cu,Ag)_{10}(Zn,Fe)_2(Sb,As)_4S_{13}$. The structure of tetrahedrite is illustrated in Figure 3.22 as an example of the structural complexity of sulphosalts.

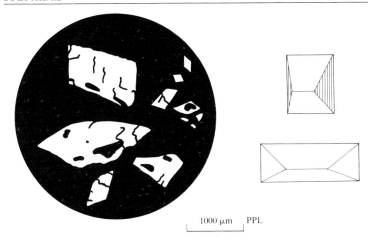

$1000\ \mu m$ | PPL

Figure 3.23 Rhomb-shaped arsenopyrite grains.

Useful reviews on sulphide mineralogy are given by Vaughan & Craig (1978), Ribbe (1974) and Nickless (1968).

Arsenopyrite (mispickel)

FeAsS

Arsenopyrite is commonly non-stoichiometric and may have Fe replaced by Co. The name "mispickel" is no longer used for arsenopyrite.

Crystals   Pseudo-orthorhombic (monoclinic) with axial ratios $a:b:c = 1.6833:1:1.1400$. Crystals are commonly prismatic [001] with twinning on {100} and {001} giving pseudo-orthorhombic crystals; {101} giving penetration twins; or {012} giving cruciform twins. Cleavage {101} is distinct. $D = 6.1$, $H = 6$.

Polished section   Arsenopyrite is white, with $R \approx 52\%$, about the same as pyrite. Bireflectance is weak but anisotropy is usually quite distinct, the colours being dark blues and browns, and extinction is poor. The anisotropy is easier to observe than that of pyrite but weaker than that of marcasite.

Grain sections are often idiomorphic rhombs or lozenges (Fig. 3.23) or rather elongate skeletal porphyroblasts. Zonation of extinction is common, and simple or hourglass twins are frequently observed. Lamellar twinning is reported. vHN = 1048–1127.

Occurrence   Arsenopyrite is considered to be typical of relatively high-temperature hydrothermal veins, where cassiterite, wolframite, chalcopyrite, pyrrhotite and gold are common associates. It is also found in most types of sulphide deposits.

Distinguishing features   Compared with arsenopyrite, pyrite is yellowish and cubic in morphology and marcasite is much more anisotropic.

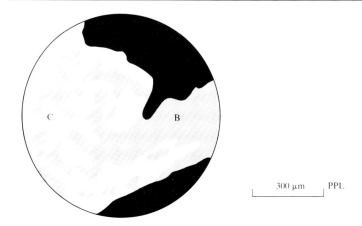

300 μm   PPL

Figure 3.24   A myrmekitic intergrowth of bornite (B) and chalcopyrite (C).

Bornite	Cu_5FeS_4
Crystals	Bornite is tetragonal (pseudo-cubic). Crystals are rare as cubes, dodecahedra or octahedra. Twinning on {111} often results in penetration twins. {111} is also a cleavage orientation. $D = 5.1$, $H = 3$.
Polished section	Bornite is pinkish brown when fresh, but soon tarnishes to purple or irridescent blue. With $R \approx 22\%$ it is brighter than sphalerite. Both bireflectance and anisotropy, with dark brown and grey tints, are very weak. Very fine granular aggregates appear isotropic. There is often a colour variation or zonation due to tarnishing. Multiple twinning is reported and cleavage traces in two directions are common. Chalcopyrite is commonly present as myrmekitic intergrowths (Fig. 3.24) or lamellae. Chalcopyrite commonly occurs along fractures. Bornite usually occurs as granular aggregates but is often intergrown with other Cu + Fe + S minerals. VHN = 97–105.
Occurrence	Bornite is usually associated with other Cu + Fe + S minerals in the "secondary environment". It can result from unmixing of high-temperature Cu + Fe + S solid solutions on cooling.
Distinguishing features	Compared with bornite, pyrrhotite is lighter brown and distinctly anisotropic; they rarely occur together.
Chalcocite (chalcosine)	Cu_2S
Digenite	Cu_9S_5

Ramdohr (1969) states that "what has hitherto been considered as 'chalcocite' with the formula Cu_2S is a great number of semi-independent minerals and solid solutions, whose relationships are not yet fully understood and for which there are diverse interpretations". Care must therefore be taken when examining samples reportedly containing chalcocite.

Crystals	Chalcocite Cu_2S is orthorhombic, $a:b:c = 0.5822:1:0.9701$. Digenite Cu_9S_5 is cubic. Both minerals are usually massive. $D = 5.77$, $H = 2\frac{1}{2}$.
Polished section	Chalcocite appears bluish light grey with $R \approx 33\%$, whereas digenite is light grey to bluish light grey, with $R \approx 21\%$. Chalcocite is weakly anisotropic, with pinkish, greenish-grey or brownish tints. Digenite is isotropic. Both minerals occur in granular aggregates, and are commonly in intergrowths with each other or other $Cu + Fe + S$ minerals. Lancet-shaped inversion twinning indicates cooling from the high-temperature hexagonal polymorph through $103\,°C$ to the orthorhombic polymorph. Cleavage traces may be observed and are enhanced on weathering. $VHN = 80-90$ (chalcocite), $80-110$ (digenite).
Occurrence	Digenite is indicative of higher temperatures and higher sulphur activity than chalcocite. Both minerals are associated with other copper and iron sulphides, especially covellite, in low-temperature hydrothermal veins and in the "secondary environment". They occur in cupriferous, red-bed sedimentary rocks and are widespread as replacement minerals.
Distinguishing features	Compared with chalcocite, djurleite $Cu_{1.96}S$ (orthorhombic) is very similar; sphalerite is slightly darker, isotropic and often shows internal reflections; and tetrahedrites are less blue, harder and isotropic.
Notes	Copper sulphide minerals are complex owing to the variation in crystallographic and optical properties with slight changes in the $Cu:S$ ratio. Their colour changes readily, owing to surface damage during polishing as well as to tarnishing.
Chalcopyrite	$CuFeS_2$ Incorporation of many other elements (e.g. Ni, Zn or Sn) is possible at high temperatures in the cubic polymorph, which has a range in composition in the $Cu + Fe + S$ system. Unmixing occurs on cooling, resulting in inclusions in chalcopyrite.
Crystals	Chalcopyrite is tetragonal, $a:c = 1:1.9705$. Crystals are commonly scalenohedral or tetrahedral in appearance (Fig. 3.25). Twinning is common on $\{112\}$ and $\{012\}$ and cleavage is $\{011\}$. $D = 4.28$, $H = 4$.
Thin section	Chalcopyrite is opaque, but alteration leads to associated blue–green staining or associated secondary hydrous copper carbonates, which are blue to green in colour.
Polished section	Chalcopyrite is yellow and it tarnishes to brownish yellow (Plate 4b). $R = 44-46\%$, slightly less than pyrite and similar to galena. Anisotropy is weak, with dark brown and greenish-grey tints, and is often not visible. Chalcopyrite usually occurs as irregular or rounded grains. It is

199

Figure 3.25
A typical
chalcopyrite
crystal

common as rounded inclusions or in fractures in other sulphides, especially pyrite and sphalerite. Colloform masses of chalcopyrite have been reported. Simple and multiple twinning is common, and cleavage traces are sometimes observed. Several phases may be present in chalcopyrite as exsolved blebs, lamellae or stars (e.g. ZnS) and indicate a high-temperature origin. VHN = 180–200.

Occurrence   Chalcopyrite is a common accessory mineral in most types of ore deposit, as well as in igneous and metamorphic rocks. It is the major primary copper mineral in porphyry copper deposits and it occurs, with bornite, in the stratiform sulphide deposits of the Copperbelt. Chalcopyrite appears to be a relatively mobile mineral in ore deposits, and commonly replaces and veins other minerals, especially pyrite.

Distinguishing features   Compared with chalcopyrite, pyrite is white, much harder and commonly idiomorphic, and gold is much brighter but may be yellower or whiter. Small isolated grains of pyrite and chalcopyrite can be very similar in appearance.

Cinnabar   HgS

Crystals   Cinnabar is trigonal, $a:c = 1:2.2905$, and it occurs as thick tabular {0001} or prismatic [10$\bar{1}$1] crystals. There is a {0001} twin plane and perfect {10$\bar{1}$1} cleavage. $D = 8.09$, $H = 2\frac{1}{2}$.

Thin section   Cinnabar is deep red. Refractive index values ($\lambda = 598$ nm) are $n_o = 2.905$ and $n_e = 3.256$.

Polished section   Cinnabar (Plates 4e & f) is light grey to bluish light grey, weakly pleochroic, with $R_o = 24\%$ and $R_e = 29\%$. Anisotropy is moderate, with greenish-grey tints, but these are often marked by abundant deep red internal reflections.

Cinnabar occurs as granular aggregates and idiomorphic crystals. Deformation multiple twinning may be present. As a result of variation in polishing hardness with orientation, granular aggregates may resemble a two-phase intergrowth at first glance.

Occurrence   Cinnabar is rare, occurring in low-temperature hydrothermal veins, impregnations and replacement deposits often associated with recent volcanics. It often replaces quartz and sulphides, and is associated with native mercury, mercurian tetrahedrite–tennantite, stibnite, pyrite and marcasite in siliceous gangue. VHN = 80–160.

Distinguishing features   Compared with cinnabar, haematite is brighter, harder and has very rare internal reflections; pyrargyrite is very similar, but with less intense internal reflections; and cuprite (Cu_2O) is bluish grey, harder and usually associated with native copper.

Notes   Metacinnabarite is a high-temperature cubic polymorph of HgS. It occurs as grains within cinnabar and is slightly darker; it is isotropic, lacks internal reflections and is softer than cinnabar.

Cobaltite   CoAsS
Cobaltite may contain significant amounts of Fe and Ni in solid solution.

Crystals   Cobaltite is orthorhombic (pseudo-cubic). It commonly occurs in cubes or pyritohedrons, but may be octahedral. There is perfect {001} cleavage. $D = 6.0–6.3$, $H = 6\frac{1}{2}$.

Polished section   Cobaltite is pinkish white with $R \approx 51\%$, slightly less than pyrite. Both bireflectance and anisotropy, with brownish to bluish tints, are weak. Cobaltite is often idiomorphic and of "cubic" morphology. It may be granular or skeletal. Colour zonation has been observed, and complex fine lamellar twinning and cleavage traces may be present. vHN = 1100–1350.

Occurrence   It is associated with Cu + Fe + S and Co + Ni + As minerals in high- to medium-temperature deposits in veins and as disseminations.

Distinguishing features   Compared with cobaltite, pyrite is yellowish and harder.

Covellite   CuS
Covelline is an alternative name recommended by the International Mineralogical Association.

Crystals   Hexagonal, $a:c = 1:1.43026$. It occurs as platy {0001} crystals with a perfect {0001} basal cleavage. $D = 4.6$, $H = 2$.

Thin section   Greenish in very thin flakes.

Polished section   Blue and strongly pleochroic, from blue to bluish light grey, except in basal sections which remain blue. $R_o = 7\%$ and $R_e = 24\%$. Anisotropy (Plate 4g, h) is very strong, with bright "fiery" orange colours.

Covellite occurs as idiomorphic platy crystals and flakes, as well as rather "micaceous" aggregates (Fig. 3.26). The good basal cleavage, parallel to the length of grains, is often deformed. vHN = 50–140.

Occurrence   Covellite commonly occurs as a "secondary" mineral after Cu + Fe + S minerals, often in the zone of secondary enrichment.

Distinguishing features   Covellite is easy to identify. Digenite is blue, but neither pleochroic nor anisotropic.

Notes   Blaubleibender covellite $Cu_{1+x}S$ occurs with covellite, and is identical in appearance except under oil immersion, when:

covellite	R_o = reddish purple	R_e = bluish grey
blaubleibender covellite	R_o = deep blue	R_e = bluish grey

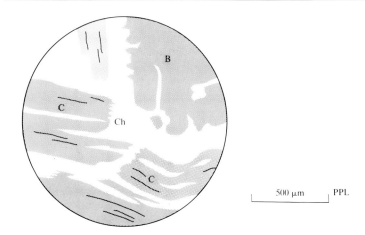

Figure 3.26 A basal section (B) and cross sections of covellite showing cleavage traces (C): chalcocite (Ch) is replacing the covellite.

Galena PbS

Galena may possibly contain some Se, Te, Ag, Sb, Bi, or As in solid solution, but usually only in trace amounts.

Crystals The crystallographic symmetry of galena is cubic and crystals are commonly cubic, cubo-octahedral and (less often) octahedral in shape (Fig. 3.27). Twinning on {111} is common, and lamellar twinning may occur on {114}. There is a perfect {001} cleavage. $D = 7.58$, $H = 2\frac{1}{2}$.

Polished section Galena is white, sometimes with a very slight bluish tint. $R = 43\%$, which makes it darker than pyrite. It is isotropic but sometimes very weakly anisotropic.

Galena commonly has cubic morphology in vein and replacement deposits. It is often interstitial to other sulphides and occurs in microfractures. Internal grain boundaries of granular aggregates are enhanced by excessive polishing. Triangular cleavage pits are characteristic of galena (Fig. 3.28 & Plate 4b) and it is often altered along cleavage traces. Many minerals occur as inclusions, but especially sulphosalts of Pb, Ag with Sb or As (Fig. 3.29). VHN = 60–100.

Occurrence Galena is common in hydrothermal vein and replacement deposits in many rock types, especially limestones. It is also common in some

Figure 3.27
Typical galena
crystals.

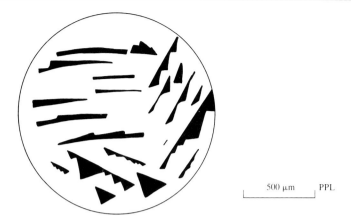

500 μm   PPL

Figure 3.28 Cleavage pits, including the characteristic triangular cleavage pits, in galena.

young (Proterozoic and Phanerozoic) stratiform massive sulphide deposits. Sphalerite is a common associate.

Distinguishing features Compared with galena, some Pb + Sb + S minerals are similar, but these are usually distinctly anisotropic.

Marcasite FeS$_2$

Crystals Marcasite is orthorhombic, $a:b:c = 0.8914:1:0.6245$. It is commonly tabular {010} but may be pyramidal. Aggregates are often globular or stalactitic. Twinning on {101} is common, often repeated, producing cockscomb pseudo-hexagonal shapes. Cleavage on {101} is distinct. $D = 4.88$, $H = 6\frac{1}{2}$.

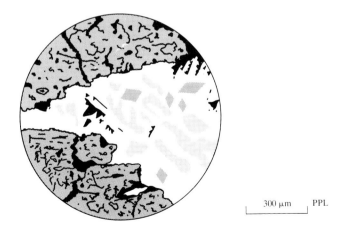

300 μm   PPL

Figure 3.29 A myrmekitic intergrowth of galena (white) and tennantite (light grey): limestone and dolomite rhombs are dark grey

203

Thin section   Marcasite is opaque, but because of ready oxidation a brown staining of limonite is often associated with it.

Polished section   Marcasite is white or slightly yellowish (Plate 4c). There is a weak pleochroism, $\|a$ pinkish white and $\|b$ and $\|c$ yellowish white. $R = 49\text{--}55\%$, very close to pyrite. The strong anisotropy of marcasite in very bright bluish and greenish greys and browns is one of the most distinctive features (Plate 4d).

Occurrence   Marcasite often appears as lath-shaped crystals in radiating aggregates of twins. Colloform aggregates with pyrite are common. Lamellar twinning and cleavage pits may be present. vhn = 901–1100.

Distinguishing features   Marcasite often occurs as concretions in sedimentary rocks. It is usually associated with pyrite in low-temperature sulphide deposits.

Compared with marcasite, pyrite is yellower, slightly softer and weakly anisotropic or isotropic; pyrrhotite is darker, brownish, softer and has a weaker anisotropy; and arsenopyrite is whiter, brighter and has a weaker anisotropy.

Molybdenite MoS_2
Molybdenite may contain Rh.

Crystals   Molybdenite is hexagonal, $a\!:\!c = 1\!:\!3.815$. Having a layer structure, it commonly has a hexagonal tabular or a short barrel-shaped prismatic habit. It is commonly foliated massive or in scales. There is a perfect basal $\{0001\}$ cleavage. $D = 4.7$, $H = 1\frac{1}{2}$.

Thin section   Molybdenite is opaque in the visible but it is transparent and uniaxial negative in the infrared.

Polished section   Molybdenite is bireflecting, with $R_o = 39\%$ (white, less bright than galena) and $R_e = 19\%$ (grey, similar to sphalerite). Anisotropy is very strong, with slightly pinkish white tints. Extinction is parallel to cleavage (the brighter R_o orientation) but is often undulatory because of deformation.

Molybdenite occurs as flakes or platelets with hexagonal basal sections. Well developed basal cleavage often results in a poor polish, especially on grains which have their cleavage parallel to the polished surface. A deformation twinning-like structure is related to buckling of the cleavage. vhn = 16–19 ($\perp$ cleavage), 21–28 ($\|$ cleavage).

Occurrence   Molybdenite is found in high-temperature hydrothermal veins and quartz pegmatites, with Bi, Te, Au, Sn and W minerals. It also occurs in porphyry copper style deposits. It is an accessory mineral in acid igneous rocks, and is occasionally a detrital mineral.

Distinguishing features   Compared with molybdenite, tungstenite WS_2 is very similar; graphite is morphologically similar but much darker; and tetradymite Bi_2Te_2S is brighter.

Notes   Molybdenite polishes poorly because of smearing.

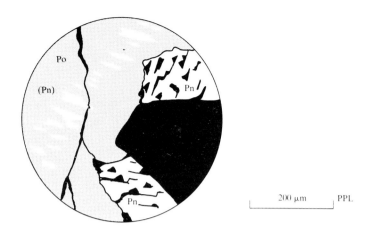

Figure 3.30 Exsolved pentlandite (Pn) "flames" in pyrrhotite (Po): also coarse pentlandite showing pits due to octahedral parting.

Pentlandite (Fe,Ni)$_9$S$_8$

Pentlandite usually contains about equal amounts of Fe and Ni. It often contains Co and sometimes Cu or Ag in solid solution.

Crystals Pentlandite is cubic, but rarely occurs as well shaped crystals. There is no cleavage, but a parting on {111}. $D = 5.0$, $H = 4$.

Polished section Pentlandite is very slightly yellowish white (cream) with $R = 47\%$. It is isotropic.

It occurs commonly as "flame" lamellae in pyrrhotite (Fig. 3.30) and as veinlets or xenomorphic grains associated with pyrrhotite. The octahedral parting {111} is often well developed, resulting in triangular cleavage pits. Alteration also takes place along this parting. VHN = 270–290.

Occurrence Pentlandite, usually associated with pyrrhotite and other Cu + Ni + Fe + S phases, is common in mafic igneous rocks, e.g. norites, and some massive sulphide deposits.

Distinguishing features Compared with pentlandite, pyrite is yellowish, often weakly anisotropic and harder, and pyrrhotite is darker, brownish, anisotropic and slightly harder.

Pyrargyrite Ag$_3$SbS$_3$

Pyrargyrite and proustite Ag$_3$AsS$_3$ are known as the "ruby silvers" because they are translucent, with a deep red colour. Extensive solid solution occurs between the two minerals.

Crystals Pyrargyrite is trigonal, $a:c = 1:0.7892$, and proustite is trigonal, $a:c = 1:0.8039$. Both minerals are commonly prismatic [0001] with twinning, sometimes complex, on {10$\bar{1}$4}. There is a distinct {10$\bar{1}$1} cleavage. $D = 5.85$ (pyrargyrite), $D = 5.57$ (proustite), $H = 2\frac{1}{2}$.

Thin section	Both minerals are deep red, and uniaxial negative.
Polished section	Both minerals are light grey, often slightly bluish. $R = 28\text{--}30\%$ (pyrargyrite) and $R = 25\text{--}28\%$ (proustite), which makes them similar in brightness to tetrahedrite. Bireflectance is distinct and anisotropy strong in greys. Red internal reflections are common, and are more abundant in proustite.

Both minerals occur as isolated crystals but are common as inclusions in galena. Simple and multiple twinning may be present. VHN: pyrargyrite 50–97 ⊥ cleavage, 97–126 ∥ cleavage; proustite VHN 70–110.

Occurrence	Pyrargyrite is more common than proustite. They are associated with other sulphosalts, especially tetrahedrite–tennantite, in low-temperature Pb + Zn mineralization and Ag + Ni + Co veins. The ruby silvers and similar Ag minerals may be significant silver carriers in base metal mineralization.
Distinguishing features	There are some rare complex sulphides which resemble the ruby silvers. Cinnabar is quite similar, but the anisotropy tints are greenish grey.

Pyrite	FeS_2

Pyrite may contain some Ni or Co. Auriferous pyrite probably contains inclusions of native gold, and cupriferous pyrite probably contains inclusions of chalcopyrite.

Crystals	Pyrite is cubic, crystals most commonly being modifications of cubes (Fig. 3.31). The {011} twin plane and [001] twin axis produce penetration twins. There is a poor {001} cleavage. $D = 5.01$, $H = 6\frac{1}{2}$.
Thin sections	Pyrite is opaque, often occurring as euhedral crystals or aggregates of small rounded grains. Alteration to limonite results in brownish or reddish coloured rims or brown staining.
Polished section	Pyrite is white, often with a slight yellowish tint, especially in small grains. $R = 54\%$, resulting in pyrite usually appearing very bright. It is only ideally isotropic in (111) sections, and the weak anisotropy in very dark green and brown can usually be seen in well polished grains.

Pyrite is usually idiomorphic but is occasionally intergrown with other sulphides, e.g. sphalerite. Grains are often cataclased (Fig. 3.32). Framboidal pyrite is common in sedimentary rocks. Growth

pyritohedron

Figure 3.31
Typical pyrite crystals.

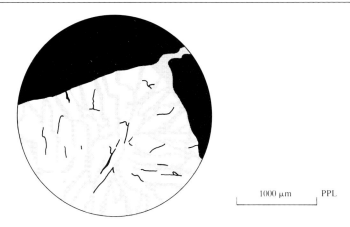

1000 μm   PPL

Figure 3.32 A cataclased pyrite cube (white) veined and replaced by chalcopyrite (grey).

zoning in pyrite is enhanced by etching. Zonation of inclusions is common. Inclusions of other sulphides, e.g. chalcopyrite or pyrrhotite, are common. Fractures in pyrite often contain introduced sulphides, e.g. chalcopyrite or galena. vHN = 1000–2000.

Occurrence   Pyrite is a common sulphide, occurring in most rocks and ores. Organic material, carbonate and quartz are all readily replaced by pyrite.

Distinguishing   Compared with pyrite, marcasite is whiter and strongly anisotropic;
features   chalcopyrite is distinctly yellow and much softer; arsenopyrite is whiter and tends to form rhomb shapes; and pentlandite is whiter, softer and often shows alteration along octahedral partings.

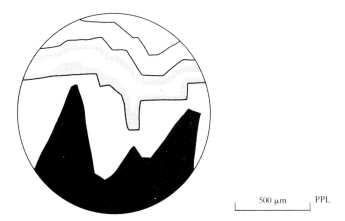

500 μm   PPL

Figure 3.33 Layers of bravoite (grey) in zoned pyrite (white) on calcite (black).

207

Notes   Melnikovite is poorly crystallized colloform iron sulphide which appears brownish and porous. It probably consists of FeS_2 and hydrous FeS. It tarnishes rapidly.

Bravoite is nickeliferous pyrite $(Fe,Ni)S_2$, often with some Co. It is similar to pyrite but brownish, slightly darker and anisotropic. It usually occurs as idiomorphic centres or as layers in zoned pyrite (Fig. 3.33).

Pyrrhotite $Fe_{1-x}S$

Pyrrhotite may contain some Ni, Co or Mn. It is cation deficient relative to the stoichiometric mineral troilite FeS. Nickeliferous pyrrhotite probably contains pentlandite.

Pyrrhotine is an alternative name recommended by the International Mineralogical Association.

Crystals   Both monoclinic and hexagonal, $a:c = 1:1.6502$, varieties of pyrrhotite occur, and these are commonly intergrown. Crystals are commonly tabular to platy with twinning on $\{10\bar{1}2\}$. There is no cleavage. $D = 4.6$, $H = 4\frac{1}{2}$.

Polished section   Pyrrhotite is brownish or pinkish white with a weak but usually visible pleochroism. $R = 35–40\%$ with R_o being darker and R_e being lighter in the case of hexagonal pyrrhotite. Anisotropy is strong, with yellowish, greenish or bluish-grey tints.

Pyrrhotite is usually xenomorphic, often occurring as polycrystalline aggregates or as inclusions in pyrite. Multiple twinning, often spindle-shaped due to deformation, is common. Exsolved lamellae (or flames) of white pentlandite are common. VHN = 370–410.

Occurrence   The presence of pyrrhotite indicates a relatively low S availability. It is common in igneous rocks, metamorphic rocks and stratiform massive Cu + Fe + S deposits. It forms by the reaction

$$FeS_2 \rightarrow Fe_{1-x}S + S \uparrow$$

in contact metamorphic aureoles. In veins, it is usually taken to indicate precipitation from relatively high-temperature, acid, reducing solutions.

Distinguishing features   Hexagonal and monoclinic pyrrhotite are not easily distinguished in polished section. A magnetic colloid may be used to stain monoclinic pyrrhotite (Craig & Vaughan 1981). Compared with pyrrhotite, ilmenite is darker and harder; bornite is browner (soon tarnishing to purple) and essentially isotropic; and chalcopyrrhotite (rare) is isotropic and browner than pyrrhotite.

Notes   Pyrrhotite alters readily along irregular fractures to a mixture of iron minerals, including marcasite, pyrite, magnetite and limonite. Although rare in sedimentary rocks and common in metamorphosed equivalents, especially near synsedimentary stratiform sulphide deposits, pyrrhotite is not thought to be necessarily a

Figure 3.34
Typical
sphalerite
crystals.

spinel-type twin

metamorphic mineral formed by breakdown of pyrite. It may be of hydrothermal exhalative origin, and could persist in seafloor sediments provided that the sulphur availability is low.

Sphalerite ZnS

Sphalerite usually contains Fe and sometimes Cd, Mn or Hg in solid solution.

Crystals Sphalerite is cubic. It has the diamond structure (see Fig. 3.22) but is more complex than one might suspect: there are many polytypes. Crystals are commonly tetrahedral and dodecahedral (Fig. 3.34). Twinning about the [111] axis leads to simple and complex twins. There is a perfect $\{011\}$ cleavage. $D = 3.9$, $H = 4$.

Thin section Pure ZnS is transparent and colourless, but sphalerite is opaque when iron rich. It has very high relief (Plate 4a) and is usually yellow to brownish in colour, with dark brown bands due to Fe zonation. Oxidation of iron-bearing varieties leads to brown staining, especially in fractures. Sphalerite is isotropic but is sometimes anomalously anisotropic, revealing fine lamellar twinning probably due to stacking polytypes. At $\lambda = 589$ nm, $n = 2.369$ (pure ZnS), 2.40 (5.46% Fe), 2.43 (10.8% Fe) and 2.47 (17.06% Fe).

Polished section Sphalerite is grey, with $R = 17\%$. It is darker than most ore minerals, but brighter than the gangue minerals (Fig. 3.35a & Plate 4b). It is isotropic. Pure ZnS has abundant internal reflections (Fig. 3.35b) but, with increasing Fe content, opacity increases and internal reflections become fewer and brownish or reddish.

Sphalerite is rarely idiomorphic. It usually occurs as rounded grains in aggregates. It also is found as zoned colloform masses. Irregular fractures are common and the cleavage often results in severe pitting. Multiple twinning is often visible. Zonation of iron, seen as brown bands in transmitted light or by internal reflection, does not visibly change brightness. Sphalerite usually contains inclusions, especially of chalcopyrite, as blebs (Fig. 3.36) or lamellae. vHN = 200–220.

Occurrence Sphalerite is common in stratabound, vein and massive sulphide deposits. Sphalerite, typically very low in Fe content, also occurs with galena, pyrite and chalcopyrite in calcareous nodules or veinlets, probably of diagenetic origin. Fe-rich sphalerite often occurs with pyrrhotite, as it is the activity of FeS rather than the

209

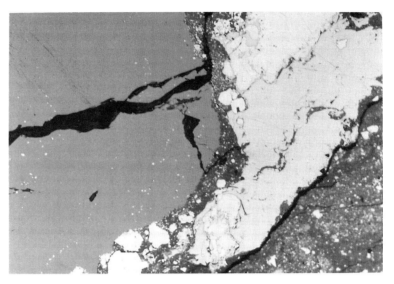

(a)

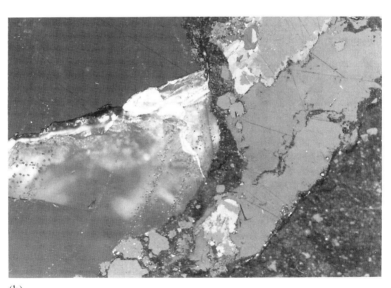

(b)

Figure 3.35 (a) A large sphalerite crystal (light grey) associated with galena (white) and marcasite (bright white), within shale (dark grey) which contains abundant grains of pyrite (white) (R-PPL). (b) Internal reflections from fractures cause the sphalerite to "glow", and thus to become translucent. Note also the anisotropy of the marcasite (top right and bottom centre). The polars are slightly uncrossed (R-XPOLS).

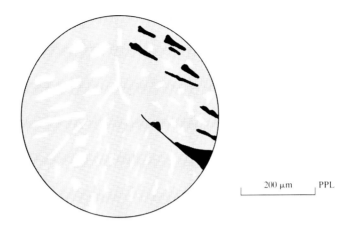

Figure 3.36 Chalcopyrite (white) blebs and interstitial veinlets in sphalerite (grey).

abundance of Fe that controls the iron content of sphalerite. Sphalerite is often associated with galena.

Distinguishing features Compared with sphalerite, magnetite is often pinkish, harder and never has internal reflections; limonite is bluish grey, usually has reddish internal reflections and is usually replacing iron-bearing minerals; and tetrahedrite is brighter, greenish or bluish grey and only very rarely shows internal reflections.

Notes Wurtzite (hexagonal ZnS) is very similar to sphalerite in polished section, but it is rare.

Stibnite Sb_2S_3

Crystals Orthorhombic, with $a:b:c = 0.9926:1:0.3393$. Crystals are usually prismatic [001], often slender to acicular. Twinning on {130} is rare. There is a perfect {010} cleavage and imperfect {100} and {110} cleavages. $D = 4.63$, $H = 2\frac{1}{2}$.

Thin section Stibnite is opaque. However, it is transparent using infrared transmitted light.

Polished section Stibnite has a pronounced bireflectance, with $R = 30–47\%$. It is light grey $\|a$, brownish light grey $\|b$ and white $\|c$. The anisotropy is very strong, with tints ranging from light bluish grey to brown. Extinction is straight.

Stibnite often occurs as acicular or bladed crystals (Fig. 3.37) or granular aggregates. The usually well developed cleavage traces are deformed, and deformation twinning is common. VHN = 70–90.

Occurrence Stibnite is found in low-temperature hydrothermal veins, usually with quartz. It is associated with complex Sb-bearing and As-bearing sulphides, pyrite, gold and mercury.

211

500 μm   PPL

Figure 3.37 Bladed stibnite grains, showing distinct bireflectance (white to grey) and cleavage traces, in quartz (black).

Distinguishing features   Compared with stibnite, haematite has a smaller bireflectance, weaker anisotropy, is harder and lacks cleavage. Some lead–antimony sulphides are very similar to stibnite.

Tetrahedrite $Cu_{10}(Zn,Fe)_2Sb_4S_{13}$

Tetrahedrite (see Fig. 3.22) exhibits extensive chemical substitution and often contains Ag, Hg and As, but only rarely Cd, Bi and Pb. The arsenic end member is tennantite $Cu_{10}(Zn,Fe)_2As_4S_{13}$. Silver-rich tetrahedrite is known as freibergite. Tetrahedrite–tennantites were often formerly called fahlerz.

Crystals   Tetrahedrite is cubic and occurs as modified tetrahedra. Twinning on the axis [111] is often repeated. There are also penetration twins. There is no cleavage. $D \approx 5.0$, $H \approx 4\frac{1}{2}$.

Thin section   Tetrahedrites are usually opaque, but iron-free and arsenic-rich varieties transmit some red light.

Polished section   Tetrahedrite is light grey, sometimes appearing slightly greenish, bluish or brownish. With $R \approx 32\%$ it is darker than galena but brighter than sphalerite. It is usually isotropic but may be weakly anisotropic. Very scarce red internal reflections have been reported from tennantite.

Tetrahedrite is rarely idiomorphic. It is usually in the form of rounded grains of polycrystalline aggregates. It forms myrmekitic intergrowths with other sulphides, e.g. galena or chalcopyrite. Zonation of Sb/As and Fe/Zn is commonly detected on microanalysis, but is not visible in polished section. Irregular fracturing is common. Inclusions, especially of chalcopyrite, are common. VHN = 240–370.

212

Occurrence   Tetrahedrite commonly occurs associated with galena in lead + zinc deposits, although it is inexplicably abundant in some and absent in others. Tennantite is common in porphyry copper mineralization.

Distinguishing features   Compared with tetrahedrites, sphalerite is darker, harder, has a good cleavage and usually shows internal reflections. Many complex sulphides (sulphosalts) are similar at first glance to tetrahedrite, but most of these are anisotropic.

Notes   The various chemical varieties of tetrahedrite cannot be identified with any certainty in polished section without resorting to microanalysis.

3.11   Tungstates

Wolframite (wolfram)

(Fe,Mn)WO$_4$

The iron end member is called ferberite and the manganese end member huebnerite.

Crystals   Wolframite is monoclinic, $a:b:c = 0.839:1:0.867$, $\beta = 90°49'$. It is usually prismatic [001]. Simple twinning is common and takes place on {100} and {023}. There is a perfect {010} cleavage and a parting on {100} and {101}.

Thin section $n_\alpha = 2.150–2.269$
$n_\beta = 2.195–2.328$
$n_\gamma = 2.283–2.444$
$\delta = 0.133–0.175$
$2V_\gamma = 60°–70°$ + ve
OAP is perpendicular to (010)
$D = 7.18–7.62$ $H = 5–5\frac{1}{2}$

COLOUR   Transparency decreases with an increase in the Fe content. Colour banding is due to a variation in the Fe:Mn ratio. Iron-rich wolframite is brownish red to dark green.

PLEOCHROISM   Common, with α red, brown or yellow, β pale green to yellowish brown, and γ red, green or dark brown.

HABIT   Elongate prismatic, often occurring as thin flat crystals.

CLEAVAGE   {010} perfect.

RELIEF   Extremely high.

*BIREFRINGENCE   Extremely high, but colours are hidden by the mineral colour.

INTERFERENCE FIGURE   Seen on a section perpendicular to (010), $2V$ is large, but it is difficult to determine the sign, since the high dispersion of the mineral makes determination colours difficult to see.

EXTINCTION   Oblique extinction, with $\gamma\hat{}cl = 17°–27°$.

Polished section   Wolframite is slightly brownish grey. With $R \simeq 16\%$ it is slightly brighter than cassiterite. Bireflectance is weak. Anisotropy is moderate and distinct in bluish greys. Extinction is oblique. Reddish-brown internal reflections are common.

Wolframite occurs as idiomorphic tabular or bladed crystals with simple twinning. Zoning is enhanced by weathering. Cleavage traces may be observed. VHN = 320–390.

Occurrence Wolframite is found in high-temperature hydrothermal veins and pegmatites, usually associated with quartz and Sn, Au and Bi minerals. It is associated with granitic rocks and greisenization. It is also found in placers with cassiterite. Scheelite $CaWO_4$ is a common associate and may replace wolframite. Scheelite is strongly fluoresecent.

Distinguishing features Compared with wolframite, cassiterite is darker and has more abundant internal reflections, while sphalerite is isotropic and is often associated with chalcopyrite.

214

4 Transmitted-light crystallography

4.1 Introduction

Light is an electromagnetic vibration which, for the purpose of transmitted- and reflected-light microscopy, can be considered as a transfer of energy by vibrating "particles" along a path from the source to the observer. White light consists of a continuous spectrum of rays, ranging in wavelength from 380 nm to 770 nm through the visible spectrum.

A ray of light defines the path by which a continuous, thin stream of energy travels outwards from a source: it represents the path that light follows in travelling from one point to another. A beam of light can be defined as a bundle of rays, and points on these rays that are – at a given moment – in the same stage of vibration are said to be in phase with each other. A surface which connects all such points in a bundle of rays constitutes a wave front or a wave surface (Fig. 4.1a). Wave normals are perpendicular to the wave front, but are not coincident with rays except in isotropic substances.

It is convenient to consider the idealized case of a single ray of monochromatic light (that is, light of a single wavelength; Fig. 4.1b). A *wave* is generated by the vibration of particles lying along the path of the ray. If the light is not polarized, the particles vibrate at random in a plane *normal* to the direction of the ray. The transverse vibrations in a ray can be considered to take place in all possible directions perpendicular to the direction of propagation. However, if the light is linearly or *plane polarized* by means of a polarizing filter, then the particles simply vibrate up and down along the line *xy* (Fig. 4.1b). A *wavelength* is the shortest distance between two points in exactly similar positions on a wave and moving in the same direction. Two waves are said to be *in phase* when they are of equal wavelength and their positions of zero amplitude occur at exactly the same time. Light of the same wavelength and of the same or different intensity (amplitude) may be either *in phase* or *out of phase*, as shown in Figures 4.1c & d. The path difference may be measured as a fraction of the wavelength. If two *coherent* rays

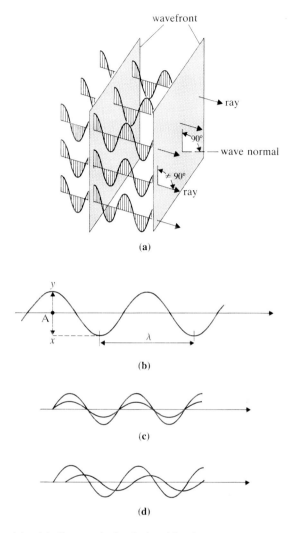

Figure 4.1 (a) Geometrical relationships between rays, wavefronts and wave normals in anisotropic media. Note that wave normals and rays are not coincident. (b) Monochromatic light (for explanation see text). (c) Two waves of the same wavelength but of different intensities, in phase. (d) Two waves of the same wavelength but of different intensities, out of phase.

(originating at the same moment from the same source) which are exactly in phase are combined, they are added together and the intensity is greatly enhanced. However, if the rays are slightly out of phase, the intensity enhancement is reduced; and, if the rays have the same amplitude and a path difference of one half of a wavelength, the vibration will be cancelled out and their combined amplitude will be zero.

In transmitted-light microscopy, plane polarized white light travels up the microscope axis, which is at right angles to the plane of the rock thin section lying on the *microscope stage*. On entering an anisotropic mineral rotated from the extinction position, the light can be considered to be separated into two components which travel with different velocities through the crystal. On leaving the crystal, the two components may be out of phase, and the *path difference* will vary for different wavelengths of light. This complexity in the light leaving the crystal is apparent only when the *analyzer* is inserted and *interference colours* are generated.

4.2 Refraction

When a ray of light strikes a surface separating one medium from another, it will be *refracted* into the second medium (assuming that the two media possess different properties) provided that:

(a) the incident ray, the refracted ray and the normal to the surface between the two media at the point of contact lie in the same plane; and
(b) the sines of the angles of the incident ray and the refracted ray, measured from the perpendicular to the surface at the point of contact, always bear a definite ratio to one another.

Under these conditions, the same ray of light will also, in part, be *reflected* back into the first medium through which it originally came, provided that the angle of incidence equals the angle of reflection.

4.3 The refractive index

The refractive index (RI) of a medium is defined as the ratio of the velocity of light in a vacuum to that in the medium. The RI varies with wavelength (a property called **dispersion**), but the variation is usually small for transparent minerals, so that only refractive indices for "white light" are usually quoted.

If V_1 and V_2 are the velocities of light in two different *isotropic* media, with i the angle of incidence and r the angle of refraction then, in Figure 4.2, $V_1/V_2 = bc/b'c' = b'c \sin i/b'c \sin r = \sin i/\sin r$ = the *refractive index*. In other words, in isotropic media the ratio between the velocities of the two media is equal to the ratio between the sine of the angle of incidence and the sine of the angle of refraction (*Snell's Law*). The refractive index (usually denoted by the letter n or abbreviated to RI) is a constant and, for a specific wave-

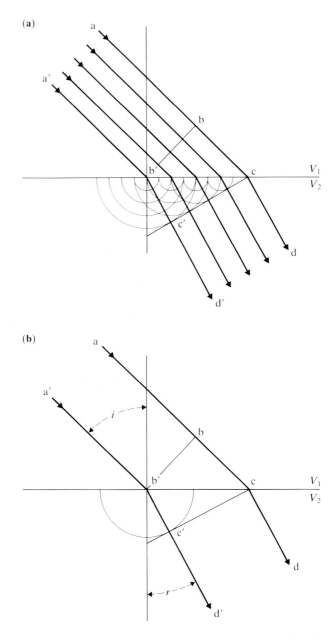

Figure 4.2 Refraction of light at a plane surface. (a) The Huygenian construction for several rays: (b) is a simplified version.

length of light, is inversely proportional to the velocity of light through the medium; that is, the RI is proportional to $1/V$. The RI usually increases as the wavelength of light decreases. Since the wavelength of red light is greater than that of blue, the refractive index of a medium in red light is usually smaller than the refractive index in blue light. White light entering a medium is split into the colours of the spectrum because the component wavelengths of the white light have different RIS; and this splitting of white light is also called *dispersion*.

4.4   The Becke effect and the Becke line

The relative refractive indices of two minerals, or one mineral and the mounting medium (either an oil of specific RI, or the cement with which a rock thin section is attached to a microscope slide), can be observed by studying the Becke effect. When rays of light strike the near-vertical contact between the two substances, some are refracted and some are totally reflected so that they are concentrated within the mineral or oil, etc. of higher RI. Under the microscope a narrow band of light – the *Becke line* – appears in this position (see Figs 1.3a & b). As the microscope objective is raised (or, in modern models, the microscope stage is lowered) the Becke line appears to move into the mineral or cement with the higher RI. The Becke line is best seen by using a higher power objective ($\times$ 10 or $\times$ 30), and stopping down some of the light passing through the mineral by means of the substage aperture diaphragm. The Becke effect is used so often that the student should observe the following rule: **As the objective lens is raised (or the stage is lowered), the Becke line moves into the substance of higher refractive index.** With minerals possessing large birefringences, such as the carbonates, the extinction position for a grain or crystal should be first obtained, so that the RI is linked to a particular optic orientation (see Section 4.8).

4.5   Birefringence and Newton's Scale of Interference Colours

If two rays, with velocities V_1 and V_2 and refractive indices n_1 and n_2, traverse a mineral plate of thickness M, in times of t_1 and t_2 respectively, then $t_1 = M/V_1 = Mn_1$ (since the velocity is inversely proportional to the RI) and, in the same way, $t_2 = Mn_2$; so that $t_2 - t_1 = M(n_2 - n_1)$. In other words, the relative retardation of the two rays is equal to the thickness multiplied by the difference in refractive indices. This latter quantity ($n_2 - n_1$) is called the **bi-**

refringence of the mineral, and is represented in the text by the symbol δ.

If, under crossed polars, white light is passed through a crystal fragment of constant thickness, a path difference of $(n/2)\lambda$ for some wavelength results, and the colour for that wavelength is transmitted. One wavelength, that of the complementary colour, has a path difference of $n\lambda$ (where n is an integer) and is removed from the white light by the analyzer; that is, it is extinguished. Other wavelengths are partly transmitted. The colour thus produced is called the **interference colour** of the mineral. Such an interference colour will not change during rotation of the microscope stage, but will only vary in intensity. The crystallographic orientation of the mineral fragment with respect to the microscope axis has an important bearing on the interference colour obtained.

If λ is the wavelength of the monochromatic light used, and P is the path difference after traversing a crystal of thickness M (in nm), then:

$$P = \text{retardation}/\lambda = M(n_2 - n_1)/\lambda$$

Thus, in a wedge of a mineral, where there is a constant difference between the refractive indices of the two components (rays) traversing the wedge, and where the thickness of the wedge varies from zero to some finite value, the path difference must increase with thickness. When such a mineral wedge (say, of quartz) is examined under crossed polars in monochromatic light, it shows alternating dark and light bands corresponding to path differences of 0λ (dark), $\frac{1}{2}\lambda$ (light), 1λ (dark), $\frac{3}{2}\lambda$ (light) and so on.

For example, if monochromatic sodium light is used, light bands are seen when the path difference is $(n/2)\lambda$, where $n = 1,3,5$, etc., and dark bands occur when the path difference is $n\lambda$, where $n = 0,1,2,3$, etc. The wavelength of sodium light (λ_{Na}) is 580 nm (where nm is nanometres, 1 nm $= 10^{-9}$ m); thus yellow bands occur at 580/2 nm, $3 \times 580/2$ nm, $5 \times 580/2$ nm etc., and dark bands at 0, 1×580 nm,, 2×580 nm, 3×580 nm, etc.

White light consists of light waves with wavelengths ranging from about 400 nm (blues) to 770 nm (reds). A quartz wedge inserted into the path of white light through the microscope produces a "spectrum" of colours. Each different wavelength gives zones of darkness and maximum light intensity for that particular wavelength at different positions along the quartz wedge (Fig. 4.3). Overlapping of the zones darkness and maximum intensity for the various wavelengths produces a series of colours, which are shown on the Michel–Levy chart on the back cover of this book. As the quartz wedge is inserted into the microscope, thin end first, the colours change from black to grey, white, yellow and then a characteristic

Figure 4.3
Quartz wedge
spectra.

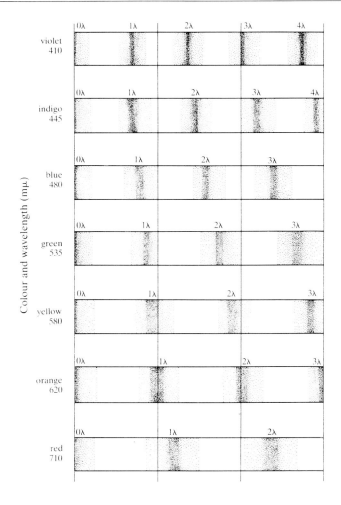

red, which marks the highest colour of the *first-order* spectrum and corresponds to a retardation of 560 nm. The *second-order* spectrum is more sharply separated into its component colours, with violet followed by indigo, blue, green, yellow, orange and finally another red. Interference effects are much more pronounced in the *third-order* spectrum and some colours are removed, the colours of this order being indigo, green–blue, yellow, red and violet. The *fourth, fifth* and *higher orders* show pale green and pinks and, at a higher birefringence a peculiar near-white colour resulting from a mixing of the component colours of white light: the interference colour seen is called "white of a higher order". It is important that the student should become familiar with the actual colours seen in the different orders, by inserting the quartz wedge into the microscope accessory slots under crossed polars, with nothing on the microscope stage, and noting the colours seen as the wedge is removed. The complete sequence of colours is called Newton's Scale of Interference Colours, but this is usually abbreviated to *Newton's Scale*, and can be observed on the *Michel–Levy chart* on the back cover of this book.

4.5.1 Anomalous interference colours

Several minerals, including chlorite and zoisite, exhibit interference colours which are *not* present on Newton's Scale, such colours being called anomalous. The most common anomalous colours are a dark blue, or Berlin blue, and a buff-coloured Berlin brown, which are seen under crossed polars. This phenomenon depends upon the dispersion of light by the minerals in question, and will not be considered further in this book, but dispersion is mentioned briefly in Section 4.3.

4.6 Isotropic and anisotropic minerals

Isotropic substances transmit light with equal velocity in all directions. A *ray velocity surface* represents the surface composed of all points reached by light travelling along all possible rays from a point source within a crystal in a given time. In isotropic crystals, the ray velocity surface is a *sphere*. Isotropic substances include glass, nearly all fluids and all minerals crystallizing in the cubic system. Another representation of the RI of a transparent medium is called an *indicatrix*. The isotropic indicatrix is also a sphere, with a radius equal to n (proportional to $1/V$, where V is the velocity of light in any direction in the substance). The ray velocity surface and the indicatrix are different but complementary representations of the RI variations for the substance.

Anisotropic crystals transmit light with different velocities in different directions, and the ray velocity surface and indicatrix of an anisotropic crystal are *ellipsoids*, which may be of two principal geometric types, *biaxial* and *uniaxial*. Anisotropic substances include all minerals crystallizing in the orthorhombic, monoclinic and triclinic systems (biaxial minerals), and also the tetragonal trigonal and hexagonal systems (uniaxial minerals).

4.7 The isotropic indicatrix

As described above, the isotropic indicatrix is a sphere in which all radii equal n (the refractive index for a specific wavelength of light), and the wave travelling along a particular radius of the sphere

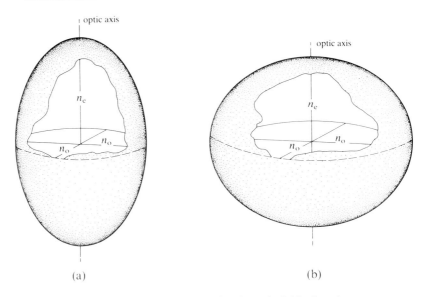

Figure 4.4 The positive (a) and negative (b) uniaxial indicatrix.

vibrates at right angles to the radius, and travels with a velocity proportional to $1/n$.

4.8 The uniaxial indicatrix

The uniaxial indicatrix applies to all minerals crystallizing in the tetragonal, trigonal and hexagonal crystal systems. In uniaxial minerals, light is polarized so as to vibrate in two mutually perpendicular planes. Light transmitted through the mineral has a velocity which depends upon its direction of propagation and its vibration direction. There is one direction along which light moves with the same velocity no matter what the vibration direction, this being parallel to the crystallographic c axis and being called the **optic axis**. Since there is only *one* optic axis, crystals with these optical properties are described as **uniaxial**.

Uniaxial crystals have two *principal* refractive indices, from which it follows that light travelling in any direction, except along the optic axis, can be considered to be resolved into two mutually perpendicular, plane polarized vibrations (components), travelling in the same direction but with different velocities. The uniaxial indicatrix (Fig. 4.4) is an imaginary surface showing the variation in the refractive indices of light waves in their directions of vibration; each radius vector represents a vibration direction and the radius length is the RI of a wave vibrating parallel to it. Waves vibrating parallel to the equatorial radii, all of which are designated n_o, are called

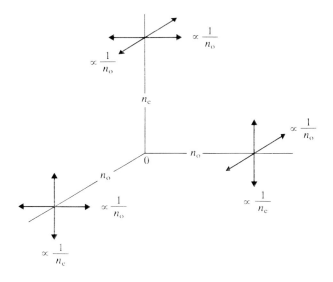

Figure 4.5 Polarization in a uniaxial crystal

ordinary or **o waves**; and rays transmitting light which vibrates parallel to the equatorial radii are called **ordinary** or **o rays**. Waves vibrating in a plane containing the optic axis have refractive indices and velocities that depend upon their direction of propagation, and are called **extraordinary** or **e waves**. The RI of a wave vibrating parallel to the c axis, designated n_e, is at a maximum or minimum value for the crystal depending upon whether $n_e > n_o$ (for positive crystals) or $n_e < n_o$ (for negative crystals). Rays transmitting light which vibrates parallel to the optic axis are called **extraordinary** or **e rays**. A wave vibrating in the plane containing the optic axis and travelling in a random direction through the crystal has an RI with a value between n_o and n_e.

Light travelling parallel to the optic axis (the crystallographic c axis) has a constant RI and a velocity proportional to $1/n_o$.

Light travelling along an equatorial radius is resolved into two mutually perpendicular components: the component vibrating parallel to the c axis has an index of n_e and a velocity proportional to $1/n_e$, whereas the other component vibrates perpendicular to the c axis, that is in the equatorial plane, with an RI of n_o and a velocity proportional to $1/n_o$ (Fig. 4.5).

Ray velocity surfaces can be constructed for positive uniaxial crystals in which $n_e > n_o$, and for negative uniaxial crystals in which $n_e < n_o$, and these are shown in Figure 4.6. In the figure, the wavefront of the ordinary ray is a sphere, whereas the wavefront of the

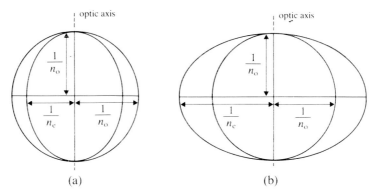

Figure 4.6 Principal sections in positive (a) and negative (b) uniaxial ray velocity surfaces.

extraordinary ray, the velocity of which varies with its direction, is an ellipsoid.

Negative uniaxial minerals include calcite, nepheline, beryl, tourmaline and idocrase. Positive uniaxial minerals include quartz, zircon and cassiterite.

4.8.1 Uniaxial crystals under crossed polars

Light entering a uniaxial mineral in any direction, *except parallel to the optic axis*, is split into two mutually perpendicular components, each travelling with different velocities; that is, one component will be relatively *fast* and the other *slow*; these two components will differ *in phase* on leaving the mineral. Consider a beam of monochromatic light of wavelength λ entering the mineral thin section from the substage polarizer, and vibrating in an east–west direction. The beam emerges from the mineral resolved into two mutually perpendicular components which differ in phase, and this phase difference is preserved until the components reach the analyzer, situated above the microscope stage, which resolves light into components vibrating in a north–south direction. The two components are combined on leaving the analyzer.

(a) If the two components differ in phase by $n\lambda$, where n is an integer, the waves combined by the analyzer are $(n/2)\lambda$ out of phase, where n is an odd number. This is because the polarizer and analyzer are orientated at $90°$ to each other. Such waves are similar in amplitude, and are in opposition whatever the orientation of the crystal section, the result being a wave of zero amplitude. In Figure 4.7, P–P' is the polarizer transmission plane, A–A' is the analyzer transmission plane, X–X' and Y–Y' are the two components into which light is resolved on passing through the crystal, OB is the amplitude of the wave leaving the polarizer, OC and OC' are the amplitudes of the two components after leaving the crystal section,

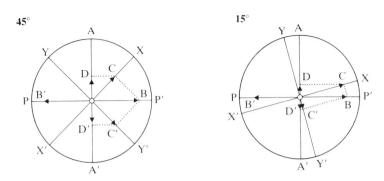

Figure 4.7 Destructive interference (for explanation see text).

and OD and OD′ are the amplitudes of these two components resolved in the analyzer. When the crystal is in the 45° position in Figure 4.7, OD and OD′ are equal and opposite, and yield a resultant wave of zero amplitude. In the 15° position, although components OC and OC′ are dissimilar, OD and OD′ are again equal and opposite, so that a wave of zero amplitude again results. Thus blackness occurs in all positions as the mineral is rotated on the stage.

(b) If the two components differ in phase by $n\lambda/2$, where n is an odd number, the components combined by the analyzer are in phase and superimposed, so that a maximum resultant wave is produced with the crystal in the 45° position, and has twice the amplitude of either of the interfering component waves. The intensity of light of this resultant wave is *four times* as great as the intensity of light of either component wave, because intensity is proportional to the square of the amplitude. This situation is illustrated in Figure 4.8, which has the same notation as Figure 4.7. In this figure, the components reinforce each other in the analyzer transmission plane.

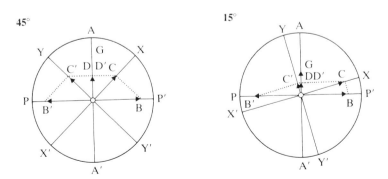

Figure 4.8 Constructive interference (for explanation see text).

In the 45° position, OD and OD′ are equal and coincident, so that the amplitude transmitted by the analyzer is 2 OD. In the 15° position, although components OC and OC′ are dissimilar, the components OD and OD′ are again equal and coincident, so that a wave of 2 OD again results. In the two situations shown in Figure 4.8, OG is equal to 2 OD. However, this condition of maximum illumination does not hold throughout a complete rotation of the microscope stage. When the two components of light produced by the mineral are parallel to the polarizer and the analyzer, that is in the 0° position – when X–X′ and Y–Y′ are coincident with P–P′ and A–A′ – the components have a value of zero, since D and D′ would be coincident with 0, and extinction results. From this it is seen that waves have a maximum amplitude (and light intensity) in the 45° position, a smaller amplitude and intensity in the 15° position, and zero amplitude (extinction) in the 0° position. Thus a mineral will extinguish four times during a complete 360° rotation of the microscope stage. Any phase difference of components from the analyzer results in a certain amount of light getting through, but the intensity of the light decreases as the path difference approaches zero.

4.8.2 Compensation and the determination of interference colour using the quartz wedge

Light passing through an anisotropic uniaxial mineral plate is resolved into two mutually perpendicular components, one *fast* and the other *slow*. Under crossed polars, the mineral can be rotated into any of the four extinction positions, where the *e* and *o* rays are parallel to the polarizer and analyzer respectively; that is, north–south and east–west. Which ray is fast depends upon whether the mineral is positive or negative. The mineral plate is rotated through 45° and the quartz wedge inserted into the appropriate slot on the microscope (usually just above the objective lens). The quartz wedge used in microscope studies *is always length-slow*; that is, the wedge is cut parallel to the prism zone of a quartz crystal. If the wedge is inserted along the slow component of the mineral, so that the slow direction of the mineral is parallel to the slow direction of the quartz wedge, it is clear that the effect is one of thickening the plate and of increased retardation, and the interference colours increase as the wedge is pushed in. This effect is called *addition*, since the retardation due to the mineral is added to the retardation due to the wedge to give a combined retardation, and hence an increase in the interference colour. The mineral is then rotated in the other direction through the extinction position and for a further 45°, so that the other component is in a position along which the quartz wedge can be inserted. This time the effect is one of decreased retardation, and hence a decrease in interference colours, until a point is reached at which the mineral interference colour is exactly "neutralized" by the wedge. At this point the combined path difference is zero, and darkness is produced on the mineral plate: this effect is called *compensation*. When darkness is obtained the mineral plate is removed, followed by slow withdrawal of the wedge, and the number of orders of interference colours seen are counted until the wedge is completely removed. This gives the interference colour for a particular mineral plate. It is usual to investigate several plates of the same mineral, and to take the highest interference colour obtained as representing the interference colour of the mineral. This technique is particularly important in minerals with high orders of interference colours, such as the olivine and epidote groups.

4.8.3 The sensitive tint plate

Apart from the quartz wedge, the most important accessory plate is the first-order red or sensitive tint plate, which is cut so that the retardation is 560 nm and the interference colour displayed by it under crossed polars is red of the first-order spectrum (first-order red). Thus a length-slow sensitive tint plate, inserted along the slow component of a mineral plate showing first-order white interference colour, will lead to the retardation of the mineral being *added* to that of the plate, resulting in a blue colour of the second order. If it is inserted along the fast component, the retardation of the mineral is *subtracted* from that of the plate, resulting in a yellow colour of the first order. Always ascertain whether the plate is length-slow or length-fast, as this is extremely important in certain microscope techniques.

4.8.4 Interference figures in uniaxial minerals and their sign

The microscope is set up as described in Chapter 1, with crossed polars, and the substage condensing accessory lens and the Bertrand lens are inserted. If the microscope does not have a Bertrand lens, the entire eyepiece should be removed. A high-power objective lens ($\times 40$ or $\times 45$) is also required.

Isotropic minerals (cubic minerals such as garnet) show no interference figures, whereas anisotropic minerals display interference figures consisting of *isogyres* and *isochromatic curves*. Isogyres are black or grey areas, which may or may not change position as the microscope stage is rotated. Isochromatic curves are colour bands, or areas which are systematically distributed with respect to the isogyres, and are seen in highly birefringent minerals.

In uniaxial crystals, an isotropic section is required to obtain a centred interference figure. This is a mineral section cut at right angles to the crystallographic c axis (that is, a section cut at right angles to the optic axis), so that the mineral grain appears black

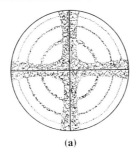
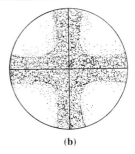

(a)   **(b)**

Figure 4.9   Uniaxial interference figures, (a) with and (b) without isochromatic curves.

under crossed polars throughout a complete 360° rotation of the stage.

The condenser lens is attached to many microscopes as a substage accessory, and is a hemispherical lens which, when inserted into the light train of the microscope, changes the inclination of the incident light to the mineral section on the stage. Thus, at the centre, light passes straight through and the crystal still behaves as an isotropic section. However, as the incident light moves further away from the centre its inclination increases until, at the edge of the lens, light is passing almost horizontally along the mineral section. Path differences occur among the components of light emerging from the crystal plate, and the *loci* of similar path differences: for example, 1λ, 2λ and 3λ appear as concentric rings of darkness, while $\frac{1}{2}\lambda$, $\frac{3}{2}\lambda$ and $\frac{5}{2}\lambda$ appear as rings of bright light for monochromatic light. This is because the substage condensing lens is hemispherical and produces "cones" of light of equal inclination. The central point, where light passes through as before, is dark, with a path difference of 0λ. When all other wavelengths are considered, the resulting **interference figure** consists of alternate concentric circles of colour corresponding to Newton's Chart. These are called **isochromatic curves**, and upon these circles a black cross is superimposed, the arms of which are parallel to the vibration directions of the polarizer and analyzer. The arms of this black cross are called **isogyres**. A uniaxial interference figure is shown, with and without isochromatic curves, in Figure 4.9.

Light passing straight through the centre of the condenser lens passes through the mineral plate travelling parallel to the optic axis, and behaves as if the crystal was isotropic (remember that we are using an isotropic section of a uniaxial mineral in this study). However, the cone of light from the outer edge of the condensing lens passes into the mineral plate almost horizontally, and is resolved into two mutually perpendicular components. Since light is travelling virtually at right angles to a basal section (the basal

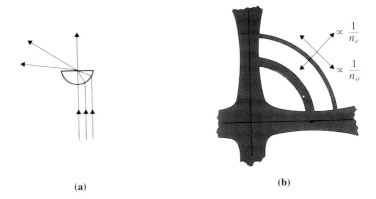

(a)   (b)

Figure 4.10 (a) Paths of light rays through the accessory condenser lens. (b) The polarisation of light in a uniaxial interference figure.

section of a uniaxial mineral is an isotropic section), it is passing into the crystal along the direction of an equatorial radius (see also Fig. 4.5). One component has an RI of n_e and a velocity proportional to $1/n_e$, whereas the other has an RI of n_o and a velocity proportional to $1/n_o$. The extraordinary component points towards the c axis, that is it points radially towards the centre of the black cross, and the ordinary component is at right angles to this, tangential to, and concentric with, the isochromatic curves (Fig. 4.10).

4.8.5   Determination of the mineral sign

After an isotropic section of the mineral being studied has been found, the microscope is set up in the conoscopic mode; that is, with crossed polars, high power objective and Bertrand lens inserted (or, if not present, with the entire eyepiece or ocular removed) and with the substage condenser accessory lens in position (or, if not present, with the aperture diaphragm open). An interference figure is then obtained and the black cross seen in the centre of the field of view; if the black cross is not centred, it is placed in the bottom left-hand corner of the field of view by rotating the stage. A length-slow sensitive tint plate is inserted so that its slow vibration direction is superimposed parallel to the component of light vibrating radially from the centre of the cross (see Fig. 4.10). If $1/n_e$ is slow (compared to $1/n_o$) for the mineral, then the retardation of the mineral and sensitive tint plate are added together, and a second-order blue colour is observed near the centre of the isogyre in the top right or northeast quadrant of the cross (see Plate 3c). The mineral is seen to be positive (+ ve). If, on the other hand, $1/n_e$ is fast (that is, in the indicatrix $n_e < n_o$), then the retardations are subtracted and a yellow colour results close to the centre of the

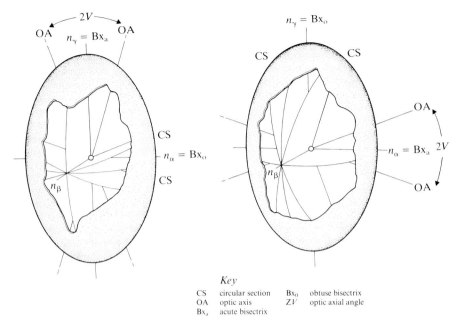

Figure 4.11 The positive (a) and negative (b) biaxial indicatrix.

cross. In this case, the mineral would be negative ($-$ve). Sign determination in a number of uniaxial figures, using either length-slow or length-fast accessory plates, and with the direction of insertion either NE–SW or NW–SE, will be described in Figure 4.16.

4.9 The biaxial indicatrix

Minerals in the orthorhombic, monoclinic and triclinic crystal systems are biaxial, and are characterized by possessing three principal refractive indices, at right angles to each other. They contain two directions normal to which light vibrates with the same velocity in all directions; that is, there are *two* optic axes (hence *biaxial*, compared with *uniaxial* crystals which have only *one* optic axis). The biaxial indicatrix is an ellipsoid, with the three major semi-axes denoted by n_α, n_β and n_γ. The three *principal sections* of the biaxial indicatrix are therefore ellipses. In the indicatrix, $n_\gamma > n_\beta > n_\alpha$, and from Figure 4.11 it can be seen that n_β is between n_α *and* n_γ in value (the length on the figure). In the principal section containing n_α and n_γ, two radial lines occur which are also exactly n_β in length. Thus two *circular sections* can be constructed, each containing one of these two radii, and the n_β semi-axis in the horizontal principal

231

Figure 4.12
Ray velocity surfaces
in biaxial crystals

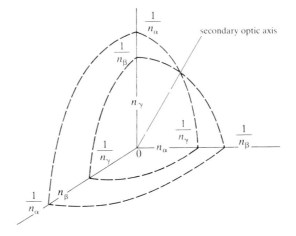

section in the diagram. The perpendiculars to *both* of these circular sections also lie in the principal section containing n_a and n_γ. Light travelling along these perpendiculars, that is at right angles to the circular sections, behaves as though the crystal were isotropic, and these directions represent the two *optic axes* of a biaxial mineral. Since these two optic axes lie on the principal section containing n_a and n_γ, this section is called the *optic axial plane* or OAP for short. The angle between the two optic axes is called the *optic axial angle*, or $2V$. The optic axial angle is bisected by a principal semi-axis, either n_a or n_γ, and this semi-axis is called the *acute bisectrix*, or Bx_a; the other semi-axis is the *obtuse bisectrix*, or Bx_o. By definition, in a positive biaxial mineral (Fig. 4.11a) Bx_a is always n_γ, whereas in a negative biaxial crystal (Fig. 4.11b) Bx_a is always n_a. Note that the third semi-axis, namely n_β, is always perpendicular to the OAP and is called the **optic normal**.

Plane polarized light travelling through a biaxial crystal is resolved into two mutually perpendicular components, as happens in *all* anisotropic minerals. Light moving parallel to the vertical semi-axis, n_γ, is resolved into the two components with velocities proportional to $1/n_a$ and $1/n_\beta$. Similarly, along the two horizontal semi-axes, the velocities of the components are proportional to $1/n_\beta$ and $1/n_\gamma$ along n_a, and to $1/n_a$ and $1/n_\gamma$ along n_β. Thus ray velocity surfaces can be constructed (as was done for the uniaxial indicatrix), and these are shown in Figure 4.12 for a positive biaxial crystal. In the complete ray velocity surface four "dimples" appear, representing the four intersection points that exist on the surface (Fig. 4.13). In all biaxial crystals, $n_\gamma > n_\beta > n_a$ and therefore $1/n_\gamma < 1/n_\beta < 1/n_a$. The difference between positive and negative biaxial crystals depends on whether n_β is nearer to n_a or n_γ in size. This dictates the position of the circular sections in the indicatrix, and whether n_a or n_γ is the acute bisectrix (Bx_a).

Figure 4.13
Ray velocity
surfaces in
biaxial crystals
in three
dimensions.

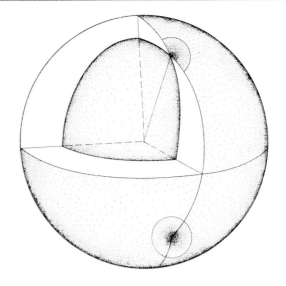

4.9.1   Interference figures in biaxial minerals

The microscope is set up in the usual mode for interference figures, as described in Chapter 1; that is, with crossed polars, the Bertrand lens in position (or, if not present, the eyepiece removed), with a high power objective ($\times 45$ or more) in place, and with either the substage condenser lens in position or the aperture diaphragm open. Two specifically orientated sections of minerals are suitable for this study, as explained below.

Mineral section perpendicular to Bx_a   The interference figure may show isochromatic curves and black brushes or isogyres. The appearance and behaviour of these on rotation depends mainly upon the $2V$. It is beyond the scope of this book to discuss the detailed evolution of the isochromatic curves, but in monochromatic light these curves are again dark and light, the dark curves representing the *loci* of points of emergence of all components of light with path differences of $n\lambda$, where n is an integer, and the light curves representing the points of emergence of components which have path differences of $(n/2)\lambda$, where n is an odd number. The dark curves are sometimes known as *Cassinian curves* (see Fig. 4.14). In white light the isochromatic curves represent the same Newton's Scale of Colours, increasing outwards from two points in the field of view (see Plate 3d). The two points mark the emergence of the two optic axes, occurring as two dark points where the path differences are zero. Remember that the field of view of a microscope in the "conoscopic mode" is such that both optic axes will be in view *only* if the optic axial angle ($2V$) of the mineral under examination is *less*

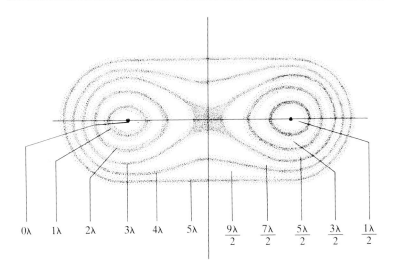

| 0λ | 1λ | 2λ | 3λ | 4λ | 5λ | $\dfrac{9\lambda}{2}$ | $\dfrac{7\lambda}{2}$ | $\dfrac{5\lambda}{2}$ | $\dfrac{3\lambda}{2}$ | $\dfrac{1\lambda}{2}$ |

Figure 4.14 A biaxial interference figure with isogyres removed, showing isochromatic curves (also called Cassinian curves).

than 45°, although this value depends upon the properties of the objective lens (see Plate 3a). Isogyres again consist of curves determined by *loci* of points of emergence of light the planes of vibrations of which are parallel to, or nearly parallel to, the planes of polarization of the polarizer and the analyzer. Isogyres appear as crosses or hyperbolae as the microscope stage is rotated (Plate 3a). The isogyres pass through the points of emergence of the two optic axes: examples of biaxial interference figures are shown in Figure 4.15 and also in Plates 3a & d. The curvature of the isogyres changes depending upon the value of $2V$; if $2V$ is large, say approaching 90°, the isogyres are virtually straight, whereas if $2V$ is small (less than 30°), they are highly curved. At one extreme, when $2V$ is nearly 0°, as in some biotites, the two isogyres touch and the interference figure appears uniaxial.

Mineral section perpendicular to a single optic axis: an isotropic section **This section is important in minerals with large optic axial angles, such as amphiboles, olivines and feldspars**. It resembles more or less half of the interference figure described above; that is, a *single* isogyre is present, sometimes with the accompanying isochromatic curves. In this type of figure any curvature of the isogyre is displayed very well and, with a little practice, it is possible to obtain a good estimate of $2V$.

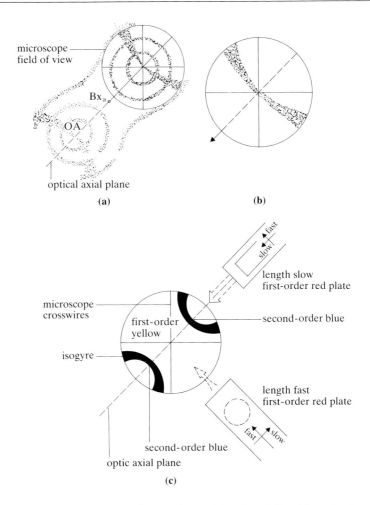

Figure 4.15 A biaxial interference figure, (a) with and (b) without isochromatic curves. The curvature of the isogyre suggests that $2V$ is large (around 60°). (c) Determination of the sign in an interference figure from a positive biaxial crystal, using either length-slow or length-fast first-order red accessory plates

4.9.2 Determination of the sign in biaxial minerals

An interference figure is obtained using one of the techniques described above. Whichever indicatrix section is employed, whether normal to Bx_a (*two* optic axes) or normal to an optic axis (*one* optic axis), the microscope stage is rotated so that the OAP (which is vertical in *both* sections) is at an angle of 45° to the microscope crosswires, and an accessory plate is inserted across it (or along it), as shown in Figure 4.15. Once again, the full optical explanation is

(a) Direction of insertion NE to SW (denoted ⤢)

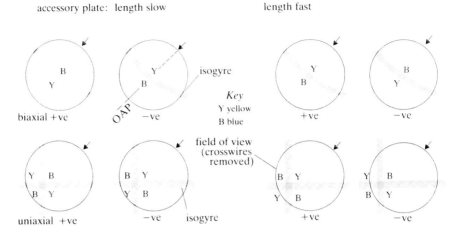

(b) Direction of insertion NW to SE (denoted ⤡)

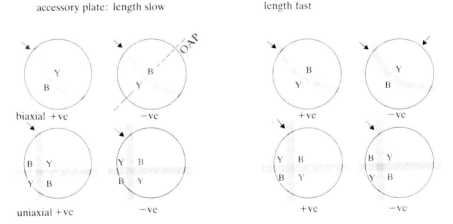

Figure 4.16 Determination of the optical sign of interference figures using either length-slow or length-fast accessory plates, and with their direction of insertion either NE–SW or NW–SE. In the sets of sketches, (a) and (d) are probably the most helpful. In (a) and (b) the uniaxial cross is placed in the lower left-hand corner of the field of view, and the biaxial single isogyre is rotated until it is concave towards the northeast. However, the cross and isogyre can also be placed in the lower right-hand corner of the field of view, with the isogyre concave towards the northwest. This explains the situation depicted in (c) and (d).

(c) Direction of insertion NE to SW (denoted ↙)

accessory plate: length slow   length fast

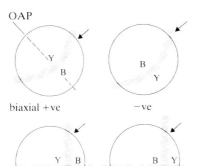

OAP

biaxial +ve   −ve   +ve   −ve

uniaxial +ve   −ve   +ve   −ve

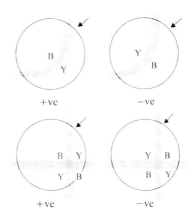

(d) Direction of insertion NW to SE (denoted ↘)

accessory plate: length slow   length fast

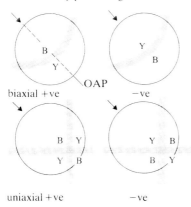

OAP

biaxial +ve   −ve   +ve   −ve

uniaxial +ve   −ve   +ve   −ve

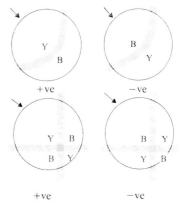

beyond the scope of this book and, although other methods involving differently orientated mineral sections can be employed, the two mineral sections already described above are the most useful and are the only ones dealt with here.

Bx$_a$ figure   In the 45° position the sensitive tint plate is inserted so that the slow direction is along the OAP. If the mineral is positive, the colour blue will appear near the centre on the concave side of the isogyre. In all biaxial minerals $n_\gamma > n_\beta > n_\alpha$ and, correspondingly, $1/n_\gamma < 1/n_\beta < 1/n_\alpha$. A positive mineral has n_γ as n_{Bx_a}, and therefore $1/n_\gamma$ is slow. Since $1/n_\gamma$ is slow, the retardation of the vibration bisecting the concave isogyre and parallel to the OAP is added to that of the sensitive tint plate, and a blue colour of the first-order spectrum of colours is obtained. A negative mineral has n_α as n_{Bx_a}, and since $1/n_\alpha$ is fast, the component vibrating parallel to the OAP is subtracted from the red colour of the sensitive tint plate, and a yellow colour indicating a negative mineral appears, again on the concave side of the isogyre (see Plate 3b).

Optic axis figure   The optic axis figure is treated in exactly the same way: the isogyre is rotated until it is pointing to the southwest, i.e. the lower left-hand corner of the field of view, and in this position the OAP is at 45° to the crosswires (running through the isogyre from SW to NE). The sign is obtained in the manner described in the previous paragraph. **When obtaining the optic sign of a biaxial mineral, always note the curvature of the isogyre and make an estimate of the size of the 2V of the mineral at the same time** (see also Fig. 4.15).

In some minerals which have 2V angles approaching 90°, it is extremely difficult to determine the curvature of the isogyre, since a single optic axis shows the isogyre as a straight black "brush" at right angles to the OAP. In such a case – as would happen, for example, with most Mg-rich olivines from basic igneous intrusions – the information obtained can merely be written as 2$V = 90°$ ± from a simple microscope study. More detailed optical information requires the use of an instrument called a **universal stage**, with which the size of the 2V can be measured directly, usually to an accuracy of ± 2°.

Sign determination in a number of biaxial figures, using either length-slow or length-fast accessory plates, and with the direction of insertion either NE–SW or NW–SE, is described in Figure 4.16.

Flash figures   Sections cut parallel to the OAP, so that n_α and n_γ lie in the plane of the mineral section, are perpendicular to n_β, and produce an interference figure known as a *flash figure*. It is possible to obtain the sign of a mineral from such a figure, but this is not recommended.

4.9.3 Determination of the extinction angle in biaxial minerals

Extinction angles in minerals are almost always measured to a prismatic cleavage or to a prism face edge. In most cases, a prismatic section (usually parallel to the c axis) of a mineral is employed in this type of study, and such a section usually shows both the trace of any prismatic cleavage(s) that the mineral possesses (appearing as frequent thin black lines), and also good anisotropic properties under crossed polars. Since a maximum extinction angle is required, it is essential to obtain a mineral section showing the greatest birefringence, since such a section will be parallel to the OAP; that is, it will have n_α and n_γ in the plane of the section, and the optic normal (n_β) will be perpendicular to the section.

The microscope is set up in the usual mode for "normal" thin-section examination; that is, with a low power objective in position, and with crossed polars. A suitable prismatic section of a mineral is obtained in the thin section under examination, and the mineral is rotated into extinction. At this point, as has been mentioned before, the two mutually perpendicular vibration components of light resolved by the mineral plate are parallel to the polarizer and analyzer of the microscope: in other words, the two light vibration components, in the extinction position, are orientated north–south and east–west – parallel to the crosswires of the microscope field of view. From previous discussion in this chapter, one of the components is *slow* and the other is *fast*. Since the section under examination is showing maximum birefringence (that is, it shows the highest Newton's Colours that can be exhibited by the mineral), the fast and slow components of the mineral in the extinction position correspond to α and γ. Inspection of the interference figure at this stage will demonstrate that a *flash figure* is seen (see Section 4.9). From the extinction position, the mineral slice is rotated through 45°, and the length-slow first-order red plate is inserted along one of the two components so as to determine whether addition (increasing the interference colour) or subtraction (decreasing the interference colour) of retardations has occurred. If, say, addition has taken place, the light component vibrating parallel to the length of the plate will be slow, and the *extinction angle will equal the angle between that component and the cleavage*. Such information is usually written as:

$$\text{extinction angle (slow or } \gamma \char`^ \text{cleavage)} = X°$$

The actual angle is measured by rotating from the extinction position to the position at which the cleavage trace is parallel to the

Table 4.1 Minerals with extinction angles that are *not* coincident with sections showing maximum birefringence.

Mineral	Section for maximum extinction angle parallel to	Components in section	Extinction angle
pigeonite	(010)	$\gamma-\beta$	$\gamma\hat{\ }\text{cl} = 37°-44°$
crossite	(010)	$\alpha-\beta$	$\beta\hat{\ }\text{cl} = 3°-21°$
katophorite	(010)	$\alpha-\beta$	$\begin{cases} \beta\hat{\ }\text{cl} = 20°-54° \\ \alpha\hat{\ }\text{cl} = 70°-36° \end{cases}$
arfvedsonite	(010)	$\alpha-\beta$	$\alpha\hat{\ }\text{cl} = 0°-30°$
kyanite	(100)	$\gamma-\beta$	$\begin{cases} \gamma\hat{\ }\text{prismatic cl} = 30° \\ \beta\hat{\ }\text{basal cl} = 30° \end{cases}$

crosswire, and then measuring the angle on the microscope stage accurately. The angle which is less than 45° is usually given. If, for example, employing the above technique, the extinction angle is found to be 51° (slow or $\gamma\hat{\ }$ cleavage), this implies that the other component (the fast one) has an extinction angle of $90° - 51° = 39°$ with the cleavage. In this case the extinction angle should be written as "fast or $\alpha\hat{\ }$ cleavage = 39°". This is termed **oblique extinction**. However, some minerals go into extinction when the two components resolved by the mineral plate are parallel and normal to the cleavage traces of the mineral (for example, the mica group minerals). In this case there is a *zero* extinction angle, and the mineral is said to possess **straight extinction**, which is shown by prismatic sections of minerals crystallizing in the orthorhombic crystal system. On the other hand, **oblique extinction** is shown by most prismatic sections of minerals crystallizing in the monoclinic and triclinic crystal systems. A mineral section showing maximum birefringence is used to obtain a maximum extinction angle for almost every biaxial mineral exhibiting oblique extinction. However, a few minerals occur in which the maximum extinction angle is not coincident with the section showing maximum birefringence, and these minerals are listed in Table 4.1. Note that the micas, although monoclinic minerals, are "pseudo-hexagonal" and show virtually straight extinction in almost every prismatic section.

4.10 Pleochroism

Pleochroism is only exhibited by *coloured minerals*, and a mineral is said to be pleochroic if it shows a change in hue, purity or intensity of colour during rotation in plane polarized light (that is, with the microscope substage polarizer in place). Pleochroism is due to the different degrees of absorption of light by the mineral in different orientations. For example, in a longitudinal section of biotite, when plane polarized light from the polarizer enters the mineral and vibrates parallel to the cleavage trace, considerable absorption of light occurs and the biotite appears dark brown in colour. If the biotite section is then rotated through 90° so that light from the polarizer enters the mineral and vibrates normal to the cleavage trace, much less absorption of light occurs and the biotite appears pale yellow.

Isotropic minerals possess the same absorption in all directions of vibration of transmitted light, so that all sections of an isotropic (cubic) mineral exhibit the same colour and are non-pleochroic.

Uniaxial minerals possess the same absorption for all light vibrating normal to the optic axis, so that sections cut at right-angles to the optic axis (basal sections) are non-pleochroic. A prismatic section is the best section in which to examine pleochroism in a coloured uniaxial mineral. In an elongate section of tourmaline, for example, the section is rotated so that its length is parallel to the north–south cross wire. If the polarizer is producing plane polarized light vibrating in an east–west plane (as is customary in many microscopes), then the light emerging from the mineral section has the colour appropriate to the ordinary ray (see Section 4.8). The mineral is then rotated through 90° until it is lying east–west, and the colour appropriate to the extraordinary ray (vibrating parallel to the crystallographic *c* axis) is seen. This gives the complete scheme for a uniaxial mineral. Thus investigation of a mineral such as tourmaline will show *o* dark brown or dark bluish green, and *e* light brown or light bluish green (see p. 151).

Biaxial minerals generally exhibit pleochroism in all sections except those cut normal to the two optic axes, since light travelling parallel to these behaves as though the mineral were isotropic. Since a biaxial indicatrix has three principal refractive indices, and consequently three principal vibration directions, a **pleochroic scheme** for a biaxial mineral can identify the three main colours appropriate to the three principal vibration directions. In order to identify each colour precisely *two* orientated sections are needed, as follows.

(a) *A section showing a centred interference figure* The interference figure is obtained in the normal way, and set up in the orientation in which the sign of the mineral is obtained. When this is done, with the curved isogyre(s) pointing to the lower left (and/or upper right) of the field of view, the OAP bisects the crosswires of the microscope, crossing the field of view from bottom left to top right. The vibration direction parallel to n_β is at right angles to the OAP (since n_β is the *optic normal*; see Sec. 4.9). The mode of the microscope is changed to normal; that is, the condenser lens and the Bertrand lens are removed, the high-power objective is replaced by a low-power objective and the polars are uncrossed (note that the stage should be *locked* while this is being done). The stage is then rotated through 45° so as to bring the vibration direction n_β into alignment with the vibration direction of the substage polarizer. The colour for this is noted (usually called β in mineral descriptions).

(b) *A section showing maximum interference colour*   Under normal microscope mode and with crossed polars, a thin section of the mineral being investigated is found that shows the maximum interference colour. In this section the *birefringence* will be at a maximum value. The maximum birefringence is equal to the difference between the maximum and minimum refractive indices of a mineral, and in a biaxial mineral this equals $n_\gamma - n_\alpha$. In such a section the OAP is lying in the plane of the section and the optic normal is parallel to the microscope axis. The interference figure obtained from this orientated section is called a **flash figure**, and the field of view tends to go from light to dark very quickly four times in a complete rotation (because the isogyres lie at right-angles to the field of view).

Having found the appropriate section, it is rotated into an extinction position, in which the two components (equal to the vibration directions parallel to n_α and n_γ) will be parallel to the directions of vibration of the polarizer and analyzer. The signs of the two components are then determined (that is, whether fast or slow) using a sensitive tint plate. A quartz wedge may be needed if the interference colours are high (see p. 227). The fast component is then identified as the vibration direction parallel to the n_α axis, and is termed α. The slow component is the vibration direction parallel to n_γ, and is called γ. The polars are uncrossed and each component is rotated in turn into the same position as the vibration direction of the light from the substage polarizer (that is east–west), and the colour appropriate to each is noted. This gives the complete pleochroic scheme for a coloured biaxial mineral.

Some minerals, such as cordierite and biotite, contain minute inclusions which may have a surrounding area that is more pleochroic than the rest of the mineral. These areas are called **pleochroic haloes**, and are due to alteration of the host mineral by radioactive emanations from the inclusions. Zircon, monazite and xenotime commonly occur as inclusions in other minerals and exhibit these haloes.

5 Reflected-light theory

5.1 Introduction

This chapter is provided to offer readers a better understanding of the theoretical basis of observations made using reflected-light microscopy, but it is emphasized that a great deal can be gained from studying ore minerals in polished section without necessarily appreciating the finer details of the theory of the optical properties of minerals. Chapter 1 provides more than enough information to get started with a reflected-light microscope. Practice and experience are probably more important in gaining confidence in the use of reflected-light than transmitted-light microscopy. However, confidence will be gained more quickly by students who have more understanding of reflected-light theory and, hopefully, many will soon find this chapter too basic, and proceed to the references given for further reading.

The nature of polarized light is described in Section 4.1. In order to understand the optical properties of minerals in reflected light, it is necessary to consider other types of polarization in addition to linearly (or plane) polarized light, which provides the basis for understanding transmitted-light optics. The polarization of light is explained in many textbooks on optics and physics, and it is discussed in detail in relation to microscopy by Galopin & Henry (1972). A brief simplified and idealized account is given here, and this should be adequate for qualitative reflected-light microscopy.

The three categories of polarized monochromatic light are illustrated in Figure 5.1, and are named according to the nature of the cross section of the wave when viewed along the path of the ray. The path of the ray is from right to left in each case in Figure 5.1, as indicated by the arrow; the cross section is projected back onto the plane which lies at right-angles to the path of the ray. Vibration of a particle up and down, to produce a wave confined to a plane, is easier to visualize than the case of a vibration which leads to ellipticity. A corkscrew is a good analogy for the special case of ellipticity which gives circularly polarized light and, if examined while it is being rotated, "waves" will be seen to travel along the length of the screw. Look down the screw from the point to see the circular cross section.

243

(a) **Line** (b) **Circle**

(c) **Ellipse**

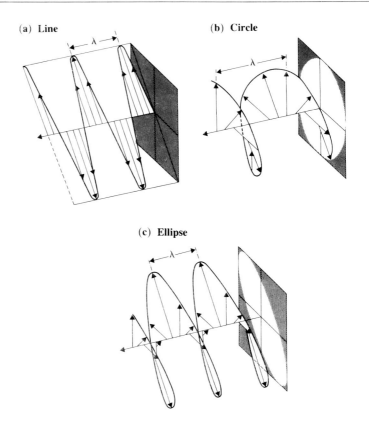

Figure 5.1 Three categories of polarized monochromatic light

Elliptically polarized light may be considered to consist of two linearly polarized components which are out of phase and vibrate at right angles; the two vibration directions correspond to the semi-axes of the ellipse shown in white in Figure 5.1c, but note that the semi-axes are not drawn in. Elliptically polarized light can be only partially extinguished (resulting in variation of the apparent brightness of the light source) by rotating a linearly polarizing filter in its path. This is in contrast to the more familiar case of linearly polarized light which is completely extinguished (cancelled, to give darkness) when its vibration direction is normal to that of the polarizing filter. In the case of circularly polarized light, the two constituent component waves have the same amplitude, but a path difference between the wavefronts of either one quarter or three-quarters of a wavelength. When a circularly polarized light source is viewed through the polarizing filter, there is no change in brightness as the filter is rotated.

In reflected-light microscopy we are dealing with *normal incident linearly polarized white light*, but the light reflected from the polished surface of a mineral remains linearly polarized only in certain cases; all sections of cubic minerals, and some sections of non-cubic minerals in certain orientations, yield reflected linearly polarized light (see Fig. 5.3). On arriving at the surface of a polished section of an anisotropic ore mineral rotated from the extinction position, the linearly polarized white light separates into two coherent components. On leaving the surface, the two components recombine, and the ratio of their amplitudes and their possible phase difference generally results in elliptically polarized light; the implication of this in relation to polarization colours is explained in Section 5.3.3. The white light reflected from ore minerals appears to the observer as "near-white" light, the brightness and colour of which depend on the optical properties of the mineral, which – in turn – relate to the chemistry and atomic structure. Sections 5.1.1 and 5.2 deal with these optical properties in more detail. The reflected light consists of a mixture of coherent rays of all wavelengths of visible light, but each wavelength may differ in intensity and azimuth, and in the nature of polarization. We can tell that the reflected light is rather complex only by inserting and rotating the analyzer and interpreting the resulting observations. This is explained in Section 5.3.

5.1.1 Reflectance

The brightness of a mineral, as observed using reflected-light microscopy, depends on factors inherent to the microscope such as the intensity of the source lamp, but it also depends on the property of minerals known as "reflectance". The *reflectance* of a polished section of a mineral is defined as the percentage of incident light reflected from the surface of the section. This reflected light travels back up through the objective of the microscope and eventually reaches the observer's eyes. Surfaces that reflect a large percentage of the light appear to be relatively bright, while those that reflect a small percentage of light of the same intensity appear to be less bright.

The reflectance value of a particular mineral is not simply a single number; it depends on variables such as the crystallographic orientation of the section through the mineral and the immersion medium used between the polished specimen and the objective. Reflectance is related to two fundamental properties of the mineral, namely the optical constants termed the "refractive index" and the "absorption coefficient". Reflectance is expressed fully by the **Fresnel equation**:

$$R\% = \frac{(n_\lambda - N_\lambda)^2 + k_\lambda^2}{(n_\lambda + N_\lambda)^2 + k_\lambda^2} \times \frac{100}{1}$$

where, for a particular wavelength λ, $R\%$ is the percentage reflectance, n is a refractive index of the mineral, k is an absorption coefficient of the mineral, and N is the refractive index of the immersion medium.

The equation holds strictly for reflection of linearly polarized light under normal incidence. It is simplified in the case of the usual observations made with "air" as the immersion medium, where $N = 1$ for all wavelengths; and in transparent minerals, which absorb a negligible amount of light, where $k = 0$.

The **dispersion** of the optical properties, i.e. their variation in value for different wavelengths of light, is much more important in understanding observations made in reflected light than in the case of transmitted light.

The **refractive index** (n) and its variation with crystallographic orientation is dealt with in the theory of optical mineralogy for transmitted-light studies (Section 4.3). However, it is worth noting that opaque minerals also have a refractive index value, even though this cannot be understood easily in terms of the velocity of light travelling *through* the section. It might be convenient to imagine that the refractive index value relates to what the velocity of light *would be* if the light were to travel through the section.

The **absorption coefficient** (k) is a measure of opacity. As light of a given wavelength passes through matter it is progressively absorbed, and the decrease in intensity is related to the absorption coefficient by the equation:

$$A = A_0 e^{-2\pi k d / \lambda_0}$$

where A_0 is the initial amplitude of a wave of wavelength λ_0, A is the amplitude after traversing a distance d in the crystal, and e is the base of natural logarithms.

The intensity of a light wave is the square of the amplitude:

$$I = A^2$$

A mineral will appear opaque when viewed using transmitted light in a thin section of standard thickness (30 μm) if its absorption coefficient is about 0.01 or greater. The absorption coefficient obeys crystallographic symmetry control, and varies with the crystallographic orientation of the section in the same way as the refractive index does. Thus, **cubic** minerals have one refractive index (n) value and one absorption coefficient (k) value. **Tetragonal**, **hexagonal** and **trigonal** minerals (*uniaxial*) have two extreme values for the ordinary and extraordinary vibration directions, $n_o \neq n_e$ and $k_o \neq k_e$. **Lower-symmetry** minerals have three end-member values, $n_\alpha < n_\beta < n_\gamma$ and $k_\alpha < k_\beta < k_\gamma$.

The optical constants can vary in value with wavelength and, following the Fresnel equation, this results in a corresponding variation in reflectance values. Hexagonal pyrrhotite (uniaxial case)

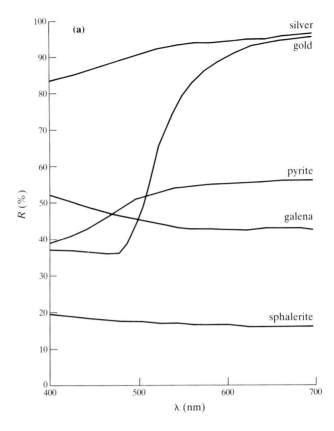

Figure 5.2(a) Examples of spectral reflectance curves for cubic ore minerals: data from Criddle & Stanley (1986).

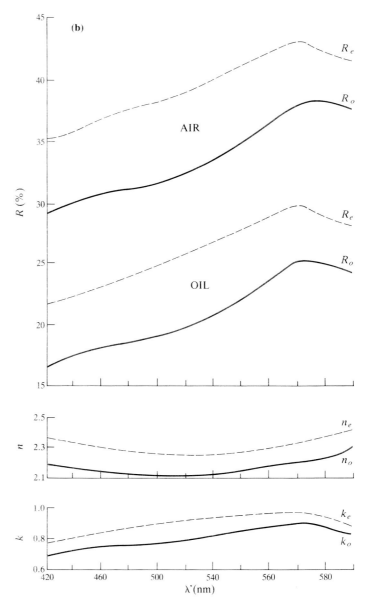

Figure 5.2(b) The variation with wavelength of R (measured in air and in oil) and the calculated optical constants, n and k, for the uniaxial mineral, hexagonal pyrrhotite (Cervelle 1979).

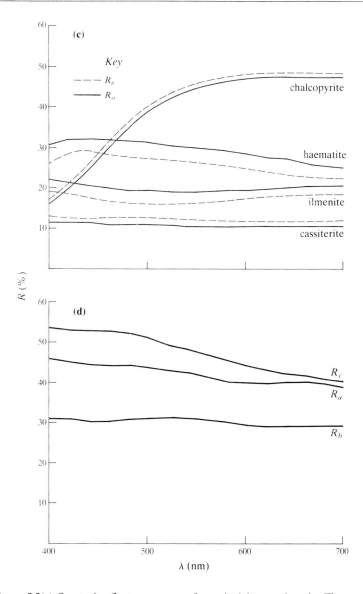

Figure 5.2(c) Spectral reflectance curves for uniaxial ore minerals. The two curves for each mineral represent reflectance for light vibrating parallel to the crystallographic c axis (R_e) and normal to the c axis (R_o): data from Criddle & Stanley (1986). (d) Spectral reflectance curves for the lower-symmetry mineral, orthorhombic stibnite. Measurements were made on sections oriented normal to the crystallographic axes. R_a for example, represents the reflectance of light vibrating parallel to the a axis: data from Criddle & Stanley (1986).

has been selected as an example (Fig. 5.2b); it can be seen how the **spectral reflectance curves** of the pyrrhotite are related to the **dispersion curves** of the optical constants. The measurement of the optical constants of opaque minerals is not as easy as the measurement of the refractive indices of transparent minerals; one method is to calculate the n and k values by solving two simultaneous equations, derived from the Fresnel equation, following measurement of the spectral reflectance values in air and in oil. The errors can be very large (Embrey & Criddle 1978) and hence reflectance values are used as the diagnostic or reference property rather than refractive indices and absorption coefficients, although the latter are more fundamental in nature. Tables of spectral reflectances have been prepared on behalf of the Commission on Ore Microscopy of the International Mineralogical Association by Criddle & Stanley (1986).

It is because of the variation in reflectance with wavelength that minerals can appear to be slightly coloured in polished section. In the case of pyrrhotite (Fig. 5.2b) the reflectance curves increase towards the red end of the spectrum, and this leads to a corresponding reddish-brown tint to the light grey of pyrrhotite. A more detailed account of colour is given in Section 5.2.

Examples of the relationship between the refractive index, the absorption coefficient and reflectance are shown for a range of minerals in Table 5.1. In these examples, the continuity in optical properties is emphasized from transparent minerals, through weakly absorbing minerals to truly opaque minerals, as the absorption coefficient increases. Referring to the Fresnel equation, it can be seen that, although the reflectance of a transparent mineral increases with refractive index, a small increase in the absorption coefficient (i.e. opacity) leads to a marked increase in reflectance.

Although it is primarily the opaque (absorbing) minerals that are studied using reflected light, the reflectance of transparent minerals can easily be calculated from the Fresnel equation. This can be helpful in characterizing "heavy minerals" in particular in polished thin section; the human eye is quite sensitive to small changes in brightness, even at the low reflectance values of transparent minerals (R less than 15%).

It is advantageous to be able to use the tabulated reference spectral reflectance curves in the *qualitative* examination of minerals in polished section. This is now possible, since the brightness, colour and pleochroism, and their variation from grain to grain, can be understood if reference is made to Figure 5.2, which shows the spectral reflectance curves for examples of cubic uniaxial and lower-symmetry minerals.

250

Table 5.1 Quantitative colour values (CIE 1931) for illuminant C (daylight) of some ore minerals. Data from Criddle & Stanley (1986). The "c" following the dominant wavelength indicates a complementary wavelength for a colour of purple tint.

Mineral		Chromaticity co-ordinates		Y (%)	λ_d	Pe (%)
		x	y			
arsenopyrite	R_a	0.315	0.321	52.5	579	2.4
	R_b	0.318	0.325	51.8	576	4.5
	R_c	0.310	0.317	51.8	498	0.2
cassiterite	R_o	0.305	0.311	10.7	476	2.4
	$R_{e'}$	0.306	0.312	12.1	476	2.0
chalcopyrite	R_o	0.349	0.369	44.1	574	24.5
	R_e	0.348	0.366	45.1	574	23.6
chromite		0.305	0.311	13.5	478	2.6
covellite	R_o	0.224	0.227	6.9	477	41.4
	R_e	0.283	0.287	23.5	476	13.4
galena		0.301	0.304	43.6	472	5.0
gold		0.386	0.388	76.1	578	39.8
hematite	R_o	0.299	0.309	29.7	483	4.7
	R_e	0.297	0.308	26.1	483	5.6
ilmenite	R_o	0.310	0.311	19.5	553c	2.2
	R_e	0.312	0.309	16.7	537c	3.1
magnetite		0.309	0.314	20.7	454	0.8
marcasite	R_1	0.319	0.329	48.6	573	5.8
	R_2	0.317	0.333	55.3	566	6.2
pyrite		0.327	0.339	53.8	573	10.6
pyrrhotite	R_1	0.330	0.334	35.3	580	10.1
(monoclinic)	R_2	0.327	0.334	40.1	577	9.3
rutile	R_o	0.298	0.303	19.7	475	5.8
	$R_{e'}$	0.301	0.306	23.0	476	4.5
silver		0.316	0.324	92.9	574	3.9
sphalerite		0.301	0.306	16.7	475	4.4
tennantite		0.300	0.312	29.6	485	4.0
tetrahedrite		0.310	0.319	32.2	541	0.6

5.1.2 Indicating surfaces of reflectance

As mentioned above, the reflectance of minerals varies with the crystallographic orientation of the section through the mineral. This directional nature of the reflectance can be described using an **indicating surface**, which is analogous, but not identical, to the refractive index indicatrix. It is a calculated surface, the calculation being made from the appropriate directional vector values of n and k. The geometrical relationship between indicating surfaces and crystal symmetry is illustrated in Figure 5.3.

The simplest surface is that for *cubic* minerals. In this case there is no variation in reflectance with orientation, so that the indicating surface is a sphere. The sphere can vary in size, with different radii (reflectance values) for different wavelengths of light, because of dispersion of the reflectance.

The surface for *uniaxial* minerals is a surface of rotation about the c axis; there can be a slight departure from a truly ellipsoidal surface, although this is usually small. The semi-axes and shape of the surface can vary with the wavelength. The reason for the departure from an ellipsoidal surface is the interaction between the refractive-index ellipsoidal indicatrix and the absorption-coefficient ellipsoidal indicatrix of the mineral, which results – via the Fresnel equation – in the non-ellipsoidal indicating surface of the reflectance.

There is no theoretically correct surface for *lower-symmetry* minerals, because only certain crystallographic orientations reflect *linearly* polarized light. Reflectance can therefore be correctly related to optical constants only via the Fresnel equation in the case of reflection of linearly polarized light from preferred vibration directions *parallel* to crystallographic planes of symmetry.

There is much less value in understanding the relationship between crystal symmetry and indicating surfaces in the case of opaque minerals than in the case of transparent minerals. This is because optic figures play an important rôle in the characterization of minerals in transmitted-light microscopy, but are relatively difficult to use in reflected light. If information is required on the use of optic figures in ore microscopy, see Cameron (1961).

5.1.3 Observing the effects of crystallographic orientation on reflectance

We are now in a position to understand observations made of reflection of light from aggregates of grains of a mineral in polished section, using the usual preliminary mode of microscopic examination, i.e. plane polarized white light. The symmetry of crystals is illustrated for reference in Figure 5.4.

Cubic minerals have one reflectance value (Fig. 5.2a) and one

253

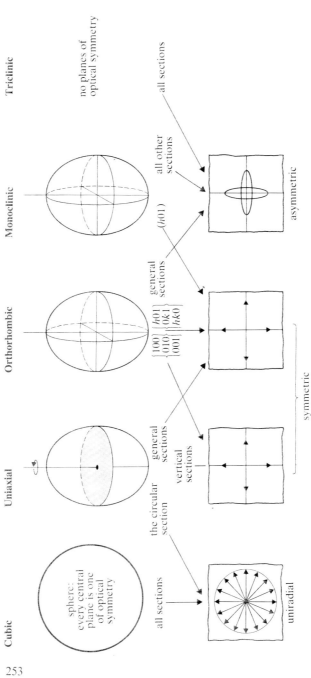

Figure 5.3 Indicating surfaces of reflectance. Considering sections in extinction positions, linearly polarized light is reflected from all sections of cubic and uniaxial minerals, but only from certain sections (those normal to a symmetry plane) of orthorhombic minerals (after Galopin & Henry 1972).

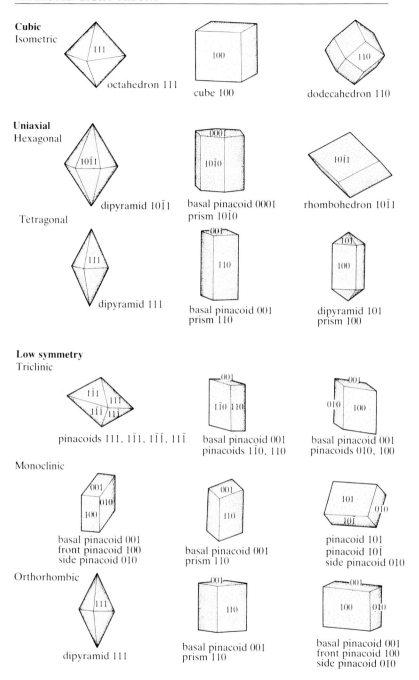

Figure 5.4 Crystal symmetry.

colour; there is no variation in brightness from grain to grain, or within a single grain, on rotation of the microscope stage.

Uniaxial minerals (Figs 5.2b & c) may vary in appearance from grain to grain, and some individual grains will vary in appearance on rotation of the stage. It should be possible at some position of the stage to make two adjacent grains of differing crystallographic orientation appear identical in brightness and colour. Sections of crystallographic orientation normal to the c axis do not vary on rotation of the stage; the reflectance value sampled in this case is that of the ordinary ray vibration direction. Sections cut parallel to the c axis show the most extreme variation on rotation, from the appearance when the incident light vibrates *normal* to the c axis (the vibration direction of the o ray), to that when the light vibrates *parallel* to the c axis (the vibration direction of the e ray). Sections of intermediate orientation will vary in appearance on rotation of the stage between two end-member cases, the first being that of the o ray, and the second being where the reflectance (and colour) is somewhere in between that of the o ray and that of the e ray. The notation e' is usually used in the latter case. Using hexagonal pyrrhotite as an example (see Fig. 5.2b) the e' spectral reflectance curve of an individual grain would fall between the two extremes corresponding to the curves labelled R_o and R_e. If a pyrrhotite crystal were carefully cut at 45° to the c axis and the surface polished, the section would be anisotropic, with four extinction positions seen in crossed polars as the stage rotates through 360°. The two vibration *directions* of the spectral reflectance curves correspond to the extinction positions. Measurement of the spectral reflectance curves in these two orientations would provide one curve corresponding to that of the o ray and one, usually labelled as $R_{e'}$, intermediate between that of the o and e rays.

In the case of *lower-symmetry* minerals (Fig. 5.2d), most grains of a mineral will vary in reflectance and perhaps colour on rotating the stage. However, some grains may be of constant appearance on rotation. Because of the complexities of the typically elliptically polarized reflected light, it is not possible to make generalizations about the behaviour of low-symmetry minerals in the way possible for uniaxial minerals.

Although the reflectance of minerals can be explained in terms of crystal symmetry, it is certainly *not* easy to determine crystal symmetry from polished sections of minerals. One reason for this is that the observed changes on rotation of the stage are often very difficult to see, especially by the inexperienced eye!

5.1.4 Identification of minerals using reflectance measurements

Reflectance measurements are made using a **spectral micro-reflectometer**. This is a special microscope adapted to permit the measurement of the intensity of reflected light from a small area of a polished section. The main components are as follows: a stabilized high-intensity light source; a monochromator, to provide "near"-monochromatic light; a diaphragm or "stop", to illuminate a small part (down to a few μm^2) of the polished grain of interest; further diaphragms, to prevent scattering of light along the optic path; a levelling device, to ensure that the polished surface is exactly normal to the optic axis of the microscope; and a photomultiplier linked to a digital measuring device, to display the light intensity value. Such microscopes have become very sophisticated, with extension of their sensitivity from the visible spectrum into the infrared and ultra-violet, and with computerized control and data-processing. A sample–standard interchange device which incorporates a method of levelling enables rapid measurements to be made at a number of wavelengths through the visible spectrum.

Reflectance values at each selected wavelength are obtained by comparison with a standard using the simple relationship:

$$R_{unknown}/R_{standard} = I_{unknown}/I_{standard}$$

where R is the percentage reflectance and I is the intensity. Aniso-tropic grains must be measured in the two extinction orientations in order to obtain the two reflectance values corresponding to each vibration direction. Measurements on sections cut and orientated using X-ray diffraction techniques are best for obtaining the extreme end-member reflectances of low-symmetry minerals when diagnostic data is being obtained for publication as a reference. Ideal standards are isotropic and have high-quality stable polished surfaces; glasses and diamond are used for low reflectance values (for example, when measuring organic matter such as vitrinite), while a cleaved surface of SiC provides a good general standard of moderate reflectance for identification of ore minerals. Pyrite is a good mineral to use as a "home-made" standard, but it requires occasional repolishing because of its tendency to tarnish.

Accurate spectral reflectance curves are now available for all of the ore minerals (Criddle & Stanley 1986). A simple identification scheme for common ore minerals was described by Bowie & Simpson (1980) and their charts incorporated reflectance values for four standard wavelengths (470, 546, 589 and 650 nm). The justi-fication in using only four values is that this is usually sufficient to characterize the main variation with wavelength of a spectral reflect-ance curve. Several grains of the mineral to be identified must be studied and reflectance measurements made at each of the wavelengths. Anisotropic grains are measured in the two extinction orientations to give the two extreme R values for the grain. It is efficient to select the grains that provide a range of degrees of anisotropy. The results are compared with the reflectance values, provided conveniently on linear charts with the background coloured appropriately for each wavelength. The variation in the reflectance data collected can provide several clues useful to identification. For example, a strongly anisotropic grain will give the approximate maximum bireflectance of the mineral, for a particular wavelength, and this should correspond to the maximum range in reflectance values seen for all of the grains. If all of the grains have one reflectance value in common, and this is true for each wavelength, then, taking into account the inherent inaccuracy of measurement, the mineral is probably uniaxial. The reason for undertaking measurements at four wavelengths is that two minerals may have similar reflectances at one of the wavelengths but different reflectances at other wavelengths, hence permitting them to be distinguished. If certain identification cannot be achieved using the reflectance values, then microhardness measurements or qualitative properties may be used to supplement the quantitative reflectance measurements. An advantage of the Bowie–Simpson method is that, although sophisticated research microscopes are required for accurate determination of spectral reflectance curves, a relatively simple apparatus can be used to provide satisfactory reflectance values at the four standard wavelengths.

The main problem with serious use of quantitative reflected-light measurements for mineral identification is that it requires too much time and expertise; it is more cost-effective to use a micro-analytical technique, such as the scanning electron microscope fitted with an energy-dispersive X-ray analyzer, following preliminary qualitative reflected-light examination. Nevertheless, there are special applications of the technique that provide a need for expertise in this area, for example, in the application of vitrinite reflectance in the determination of maturity of hydrocarbon source rocks.

Now that the reference data is available, it would be of enormous benefit to users of reflected-light microscopy if a simple method of light intensity measurement could be incorporated into basic reflected-light microscopes, in the way in which sensors have been incorporated into cameras. The estimation of the reflectance value is a key step in the identification of minerals in reflected light, but the human eye is very poor at estimating absolute levels of light intensity.

257

5.2   Colour of minerals in PPL

Recognition of the colour of minerals in polished section is very useful in their identification, but unfortunately most minerals are only slightly coloured, and the actual colours seen are easily changed. The colour change may be real. For example, it may result from slight tarnishing which tends at first to enhance the colour; then as a thin film of oxidation product forms, variable strong colours due to thin film interference effects can be produced. Some sulphides, such as marcasite, pyrrhotite and bornite, are very prone to such alteration. The colour change can also be illusory, and this often results from the effect on colour vision of a varying background colour due to different associated minerals. Tetrahedrite is light grey, but the colour tint of tetrahedrite inclusions can vary in appearance from bluish to greenish, depending on the colour of the host phase.

The application of *quantitative colour theory* to ore minerals has led to a better understanding of colour and its use in mineral identification. The colour perceived by the observer depends on the nature of the light source, the spectral reflectance curve of the mineral and, finally, on how the brain interprets the spectral distribution of the light reaching the eye in terms of the mineral's surroundings. It is this final step that causes most problems for beginners. It must be emphasized that, as in the case of brightness, the colour of some minerals can appear to be slightly different under different circumstances, and viewers with different colour "experience" may see or express the colour in slightly different ways. The possibility of slight imperfection in the observer's colour vision must be considered and, obviously, anyone who is severely colour blind will have much more difficulty in using reflected-light microscopy.

The advantage of the quantitative colour approach is that it should be possible to overcome the inherent weaknesses in human vision by providing a numerical representation of colour obtained by some measurement process. The quantitative colour system adopted for microscopy is that of the Commission Internationale d'Eclairage in 1931 (see Judd 1952). This system has been used extensively in industry to provide colour values for paint, paper, cloth and even vegetables. More recently, it has formed a basis for calculating colour in computer graphics applications. The only measurement required to obtain the quantitative colour values of a polished section of a mineral is its spectral reflectance curve. This curve represents the modification made by the polished surface of the mineral to the white colour of the light source. A hypothetical surface with 100% reflectance at all visible wavelengths would obviously appear bright white (the colour of the source lamp). All minerals have reflectances much less than 100%, and since $R\%$ varies with wavelength this leads to colour – but not in a simple way. The human eye samples the light in terms of its red, green and blue (RGB) components, and sampling the spectral reflectance curve with the spectral sensitivity of the human eye for RGB is the basis of the calculation of the chromaticity co-ordinates which are used to plot the colour on the CIE (1931) colour diagram (described in Section 5.2.1). Minerals can be plotted on this diagram, and their colours compared quantitatively as well as qualitatively.

Quantitative colour values of ore minerals (see Table 5.1) are readily available in the IMA/COM DATA FILE (1977) and in the database of Criddle & Stanley (1986). They are presented as three numbers: *visual brightness* ($Y\%$) corresponding approximately to the reflectance in white light; *dominant wavelength* (λ_d), which indicates the hue of the colour; and *saturation* ($P_e\%$), which indicates the strength of the colour. Thus bright white with a slight greenish tint would correspond to $Y\% = 50$, $\lambda_d = 585$ and $Pe\% = 1$, while bright green corresponds to $Y\% = 45$, $\lambda_d = 585$ and $Pe\% = 30$ (refer to Fig. 5.5). The colour values of a mineral depend on the type of light source; only the A (tungsten light) and C (daylight) sources need be considered for reflected-light microscopy. Note that the computation of the colour for a particular source is independent of the source used to determine the spectral reflectance curve.

Cubic minerals have one reflectance curve (Fig. 5.2a) and therefore only one colour. A *non-cubic* mineral has a colour corresponding to each of its spectral reflectance curves (Figs 5.2b–d); a typical section orientated at random through the crystallographic axes of the mineral would be pleochroic, but the pleochroism is usually very weak. Some sections of non-cubic minerals are isotropic and therefore non-pleochroic. Bireflection and pleochroism are closely related properties, and a change in brightness influences the colour perceived. However, as far as possible, the term "bireflection" should be used when the only change is in brightness, whereas "pleochroism" should be used when the change is in dominant wavelength or purity. Simple colour terminology, e.g. bluish white (not pale lavender blue!), should be used in mineral description, keeping in mind the three RGB parameters used to quantify colour.

Since colour values carry all the information of the spectral reflectance curve, it is possible to identify minerals using quantitative colour rather than by simply obtaining reflectance values. The colour of the mineral can be determined using special comparison microscopes, in which a reference colour can be changed using filters until it matches the colour of the mineral. Alternatively, spectral reflectance measurements can be used; an ore mineral identification scheme (NISOMI-81), based on quantitative colour

measurements obtained using a spectral micro-reflectometer inter-
faced to a computer, has been developed and described by Atkin &
Harvey (1982).

The use of quantitative colour is one stage more sophisticated than
the use of reflectance measurements to identify ore minerals, and even
more expertise and expense is involved. Other analytical techniques
are therefore usually quicker and more cost-effective, as mentioned in
Section 5.1.4. However, a knowledge of quantitative colour theory is
useful, and it is important to emphasize that established quantitative
colour values can be used as an aid to mineral identification without
the need for the observer to undertake spectral reflectance measurements. Such use of quantitative colour values will soon be appreciated if the exercise in Section 5.2.2 is studied.

5.2.1 CIE (1931) colour diagram

All colours visible to the human eye under certain conditions plot in
the colour diagram of Figure 5.5, within the field enclosed by the
spectral locus, which is the curve extending from 380 to 770 nm, and
the **purple line** which joins the two ends of the curve. This area is
two-dimensional in terms of colour, but brightness can be plotted as
a vertical axis, normal to the plane of the paper, and it gives a
three-dimensional mountain-like body, with the peak corresponding to 100% brightness (pure white) at point C (the white colour of
the source light); the perimeter around the base of the mountain
corresponds to 0% brightness (blackness due to no light intensity).
Colours plot within this mountain, but ore mineral reflected-light
colours tend to plot in a restricted zone from bluish through white
to yellowish; there are only a few minerals with green or purple–red
tints. As most minerals are only slightly coloured, they plot close to
point C. Covellite (basal section, o ray) is plotted as an example; it is
the "deepest" blue mineral. Its approximate quantitative colour
values (for the C illuminant) are:

Covellite (R_o): chromaticity co-ordinates $x = 0.224$, $y = 0.226$
dominant wavelength   = 475 nm
purity   = 42%
brightness (Y)   = 6.8%

The dominant wavelength is given by a projection of a line from C
through the colour of the mineral (e.g. "Cov" in Fig. 5.5) to the
spectral locus, while the percentage purity is given by the closeness
of the plotted point, "Cov", to the spectral lcous, i.e.
$a/(a + b) \times 100$.

This figure may be copied in an enlarged form and used as a basis
for the plotting of colour values from the standard references, or to
undertake the exercise given in the following section.

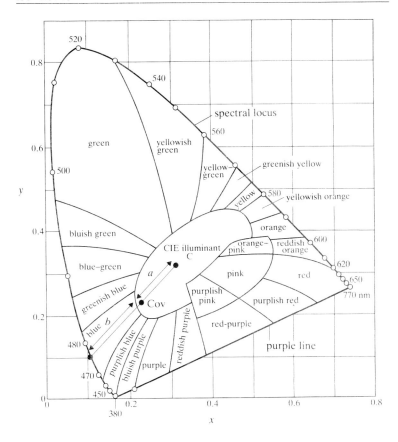

Figure 5.5 The CIE (1931) colour diagram, with colour areas (Judd 1952).

5.2.2 Exercise on quantitative colour values

Chromaticity co-ordinates and the visual brightness ($Y\%$) of an unknown mineral (**B**), sphalerite and the basal section of covellite for an A source (tungsten lamp) are given below, and plotted on the CIE colour diagram shown in Figure 5.6.

Plot mineral B on the diagram and explain, using quantitative colour values, how this mineral would appear in polished section. (Answer given at end of chapter.)

5.3 Isotropic and anisotropic sections

5.3.1 Isotropic sections

Isotropic sections appear dark, ideally black, using crossed polars and they should not change in brightness on rotation of the stage. They will appear brighter and perhaps coloured if the analyzer is

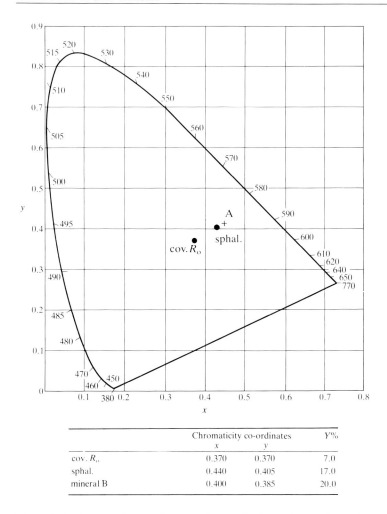

	Chromaticity co-ordinates		$Y\%$
	x	y	
cov. R_0	0.370	0.370	7.0
sphal.	0.440	0.405	17.0
mineral B	0.400	0.385	20.0

Figure 5.6 An exercise on the use of quantitative colour values: the CIE colour diagram for an A source.

slightly rotated, but again there should be no change in the appearance of the mineral section as the stage is rotated. If all grains, i.e. small sections in different crystallographic orientation through a mineral, appear to be isotropic, then the mineral is probably cubic. However, the mineral could be non-cubic, but with a very weak, imperceptible, anisotropy. Basal sections of uniaxial minerals and some sections of lower-symmetry minerals are isotropic. Sections which appear to be isotropic, but not black, may have a component of circularly polarized reflected light.

The distinction between isotropic and anisotropic sections or minerals is not as easily made (nor is it as significant) in the case of reflected light as in the case of transmitted light.

5.3.2 Anisotropic sections

Anisotropic sections show colours, known as polarization or aniso-tropic colours, using crossed polars. The colour effects are usually weak, e.g. dark reddish-brown, dark bluish-grey. Anisotropy is best detected using slightly uncrossed polars, but it must be remembered that this may change the tint of the polarization colour. Some of the grains of a non-cubic mineral will have stronger anisotropy than others, and some may be isotropic. Some cubic minerals may exhibit anisotropy; in the case of pyrite this is probably related to the relatively low symmetry within the cubic class.

Using exactly crossed polars, general sections of uniaxial miner-als, i.e. sections oblique to the *c* axis, have four extinction positions during 360° rotation of the stage. The extinction positions occur every 90°, and the colours exhibited in each 45° quadrant should be identical. However, even very slight misalignment of the polarizer or analyzer, or strain in the objective, may change the colours slightly. For this reason, the actual colours seen must be treated with caution when using tabulated reference data for mineral identification. Lower-symmetry minerals also show polarization colours, but they need not have distinct extinction positions, nor show the same colours in each 45° quadrant.

5.3.3 Polarization colours

Polarization colours differ in origin from interference colours, which are seen using transmitted-light microscopy. Their origin can be explained with the help of Figure 5.7, which illustrates the simple case of reflection from a **uniaxial transparent** mineral, such as calcite, in the 45° orientation, i.e. rotated 45° from the extinction position.

Normal incident linearly polarized **monochromatic** light, vibrating east–west, is resolved into two components, the two vibration direct-ions (corresponding to the extinction orientation) on the surface of the section. On leaving the surface, recombination of the components results in reflected linearly polarized light, vibrating in a direction closer to the vibration direction of the higher reflectance. The reflected light is now no longer vibrating normal to the analyzer, and some of it will be able to pass through the analyzer. The *smaller* the angle between the new vibration direction of the light and the vib-ration direction of the analyzer, the *greater* the amount of light that will pass through the analyzer. Obviously, the *larger* the difference is between R_{max} and R_{min} the *larger* the **angle of rotation** will be and this will result in more light passing through the analyzer.

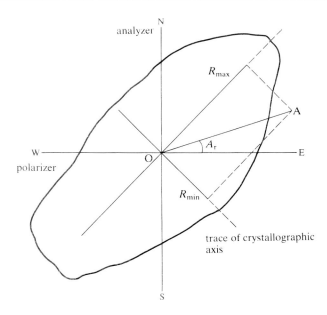

Figure 5.7 The geometry of reflection of normal incident linearly polarized monochromatic light from a bireflecting (*hkl*) grain of a uniaxial transparent mineral, turned to 45° from extinction. The incident light vibrating in the plane of the polarizer (east–west) is resolved, on the polished surface, along the two principal axes R_{max} and R_{min}, corresponding to n_{max} and n_{min}. This results in rotation of the plane of polarization (i.e. the azimuth of vibration) through the angle A_r, so that the reflected light is linearly polarized but vibrating parallel to OA.

The angle of rotation may be dispersed, i.e. it may vary with wavelength, because the reflectance values corresponding to the two vibration directions of the crystal section may be dispersed; the *amount* of light of *each* wavelength passing through the analyzer can therefore vary and this produces a colour, the polarization colour, when incident white light is used. The colours are usually weak, because most of the light is cut out by the analyzer.

Further complications arise when considering **opaque** (absorbing) uniaxial minerals. Because of the different absorption coefficients (k) of the two vibration directions of the crystal section, the reflected light is no longer linearly polarized, but is elliptically polarized. The **ellipticity** results because the two components that combine differ in both amplitude *and* phase. It is the difference in absorption which "slows down" one component relative to the other and produces a phase difference. Some of the reflected light will pass through the analyzer because of ellipticity as well as rotation. Dispersion of the degree of ellipticity contributes to colour effects.

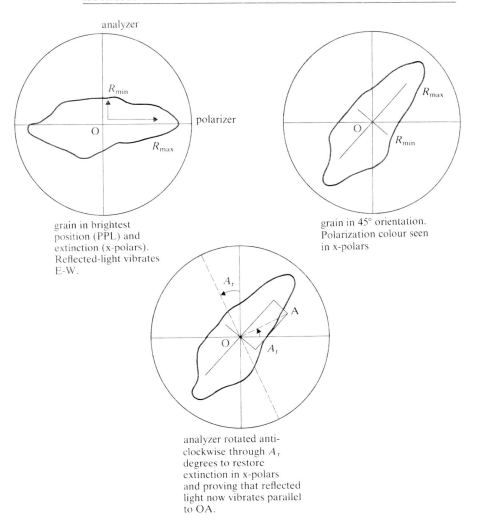

Figure 5.8 An exercise to demonstrate the rotation of polarized light on reflection from an anisotropic grain, e.g. ilmenite or hematite.

So far, only the uniaxial case has been considered. The theory of reflection as outlined above for uniaxial minerals applies only to lower-symmetry minerals for sections *normal* to a crystallographic symmetry plane; only these sections, when in the extinction orientation, will contain two vibration directions which reflect light that remains *linearly* polarized. General sections, i.e. sections *oblique* to symmetry axes, produce *elliptically* polarized light. Because of the differing crystallographic orientation of the three end-member refractive indices and the three absorption coefficients, and the

265

dispersion of their orientation, the origin and behaviour of polarization colours on rotation of sections cannot be represented in a simple way for low-symmetry absorbing minerals. Also, the concept of optic axes and isotropic sections of biaxial transparent minerals does not have a simple analogy in the case of absorbing minerals (see Galopin & Henry 1972, p. 88).

5.3.4 Exercise to demonstrate rotation on reflection

This exercise, illustrated in Figure 5.8, provides a method of demonstrating the rotation of polarized light on reflection from an anisotropic grain. Ilmenite or haematite, which are uniaxial, are suitable, but any white or light grey distinctly anisotropic mineral could be used, although the experiment may be less satisfactory for lower-symmetry minerals.

(1) Select a single optically homogeneous elongated grain of reasonable size, which exhibits distinct bireflectance and distinct anisotropy. Sketch the grain, viewed in PPL, but positioned in one of the four extinction positions, which must be located using exactly crossed polars. Indicate the reflectance values R_{max} and R_{min} of the two vibration directions on the sketch. If the grain is distinctly bireflecting it should be relatively easy to determine whether R_{max} lies east–west (grain at its brightest in PPL) or R_{min} lies east–west (grain at its darkest in PPL).

(2) Rotate the grain exactly 45° from extinction so that the R_{max} vibration direction is aligned NE–SW. Sketch the grain, again showing R_{max} and R_{min}, using a longer line for R_{max} to signify the greater percentage of light reflected. If you know, or can estimate, the values of the two reflectances, then the lines can be drawn to scale, but this is unnecessary for the purpose of the exercise.

(3) On your sketch, complete the rectangle to show the vibration direction of the reflected light (OA in Fig. 5.8).

(4) To prove that the light is in fact vibrating in this direction, push in the analyzer and rotate it very slowly a few degrees counterclockwise (or rotate the polarizer clockwise from 0° towards 90°) until a position of darkness is obtained. Adjust the rotation slightly until the maximum darkness is obtained. This rotation of the analyzer causes its vibration direction to become normal to that (OA) of the reflected light, so resulting in extinction. Rotation of the reflected light from east–west to OA on reflection from the polished surface of the mineral is therefore confirmed. The renewed extinction will not be perfect because the reflected light, vibrating in the direction OA, is

elliptically polarized. The "apparent angle of rotation", A_r, cannot be measured with sufficient accuracy to be of much assistance in identification using typical student microscopes.

If there is some uncertainty in determining the orientation of R_{max}, then the above procedure can be followed and will be successful if, by chance, R_{max} is oriented NE–SW. However, should R_{min} be oriented NE–SW then, instead of obtaining a position of darkness after rotation of the analyzer through a few degrees, the polarization colour will continue to become brighter. In fact, this technique can be used to *determine* the R_{max} and R_{min} orientations.

Note that minerals which show strong pleochroism in PPL and/or vivid polarization colours (e.g. covellite) display *dispersion* of the angle of rotation. Once the grain is in the 45° orientation, rotation of the analyzer produces variable colours rather than a simple position of maximum darkness. This display of colours is explained by extinction of some wavelengths at a given angle of rotation, while others are transmitted to varying degrees. Slight movement of the analyzer changes the distribution of extinguished and transmitted wavelengths, and produces a different colour. In a simple example, a blue colour would result from extinction of red and vice versa (Plates 4g, h).

5.3.5 Detailed observation of anisotropy

When viewing an anisotropic grain using exactly crossed polars, the "strength" of the anisotropy may be estimated from the amount of light reaching the eye, i.e. the intensity of the image, when the grain is in or close to the 45° orientation. In the case of a **strongly anisotropic** mineral, the anisotropy will be immediately evident when a group of grains of the mineral are examined and the stage rotated. The grain showing the strongest anisotropy can then be examined more closely to obtain further information. The term **weakly anisotropic** can be used if the anistropy is only visible with strong illumination and with the use of slightly uncrossed polars.

When determining the actual colour tints of the polarization colours, these should at first be observed using *exactly* crossed polars, so that the effect of *uncrossing* the polars can be noted. The vividness of the colours, i.e. how colourful they are, is an indication of the dispersion of the rotation angle and degree of ellipticity. Two examples may help in explanation: a bright grey colour represents strong anistropy but small dispersion; and a dark blue colour represents weak anisotropy but large (or strong) dispersion. "Distinct anisotropy" is a useful term, because it simply indicates how easy it is to see the anisotropy.

Published reference data would be expected to give the polarization colours of the mineral as obtained using exactly crossed polars. However, the polarization colours observed using slightly uncrossed polars will usually be sufficiently characteristic of the mineral to be useful in identification.

A mineral showing four good extinction positions at 90° and the same tint at 45° either side of an extinction position is probably uniaxial. If most sections show poor extinction and the polarization colours cannot be balanced about the "best" extinction position (the position of maximum darkness), the mineral is probably of lower symmetry.

The eye is best trained in the study of anisotropy by examining polished minerals of known identity and of varying anisotropy, and by comparing observations with those given in reference tables.

Answer to problem in Section 5.2.2   Plot the mineral B onto the diagram using its chromaticity co-ordinates x and y. Draw a straight line from A through B to the spectral locus. All three minerals should lie on this line, and they have a dominant wavelength of 486 ± 4 nm. This means that B is bluish in colour and the hue (shade or tint) of blue is exactly the same as that of covellite. The distance of a mineral from A towards the spectral locus indicates the purity (saturation or depth) of the colour. As sphalerite is essentially colourless and covellite is distinctly blue, we can say that B will be slightly bluish. The $Y\%$ values (brightness) approximate to $R\%$ for white light. Covellite is quite dark: sphalerite is grey, brighter than typical gangue minerals such as quartz, but darker than the common and more absorbing ore minerals such as haematite. Mineral B has a reflectance slightly greater than that of sphalerite.

In summary, mineral B will appear in polished section as a slightly bluish light-grey mineral; it will be slightly brighter than sphalerite, and the blue colour will be of the same hue as the covellite basal section. This *comparative* description would help experienced microscopists to visualize the problem mineral B in relation to the common minerals, sphalerite and covellite.

Appendix A   Refractive indices

A.1   Biaxial minerals

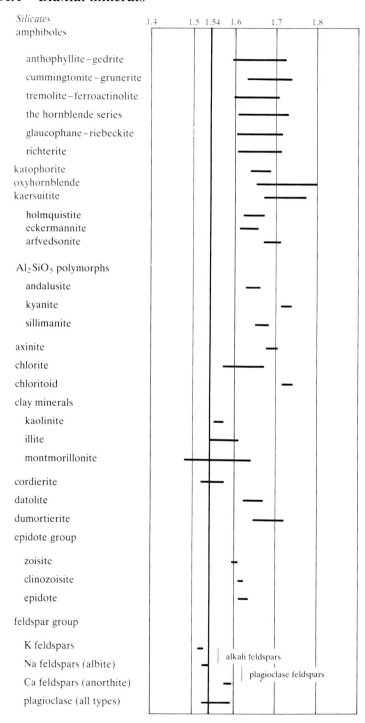

Silicates
amphiboles

anthophyllite – gedrite
cummingtonite – grunerite
tremolite – ferroactinolite
the hornblende series
glaucophane – riebeckite
richterite
katophorite
oxyhornblende
kaersuitite

holmquistite
eckermannite
arfvedsonite

Al_2SiO_5 polymorphs

andalusite
kyanite
sillimanite

axinite
chlorite
chloritoid
clay minerals

kaolinite
illite
montmorillonite

cordierite
datolite
dumortierite
epidote group

zoisite
clinozoisite
epidote

feldspar group

K feldspars
Na feldspars (albite)
Ca feldspars (anorthite)
plagioclase (all types)

alkali feldspars
plagioclase feldspars

1.4   1.5 1.54 1.6   1.7   1.8

270

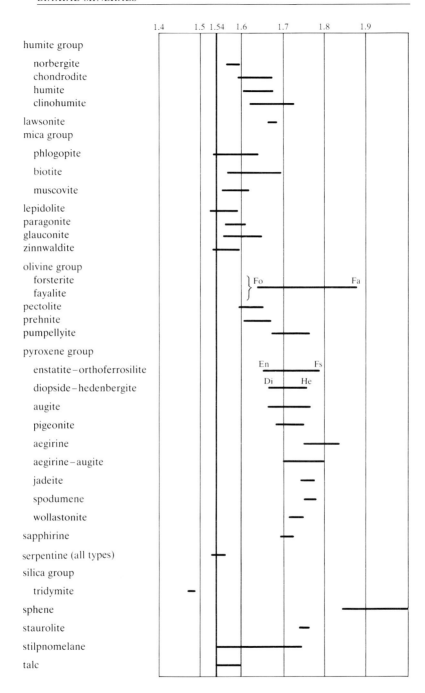

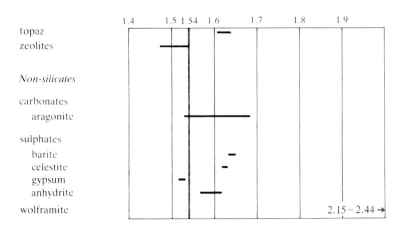

topaz
zeolites

Non-silicates

carbonates
  aragonite

sulphates
  barite
  celestite
  gypsum
  anhydrite
wolframite

A.2   Uniaxial positive minerals

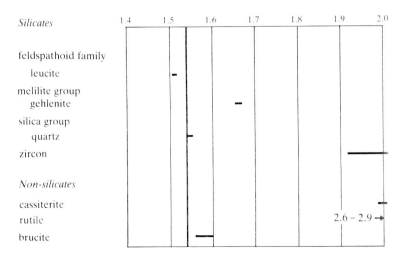

Silicates

feldspathoid family
  leucite
melilite group
  gehlenite
silica group
  quartz
zircon

Non-silicates

cassiterite
rutile
brucite

A.3 Uniaxial negative minerals

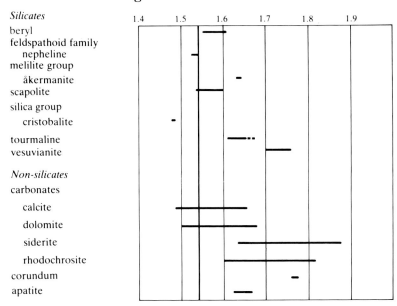

A.4 Isotropic minerals

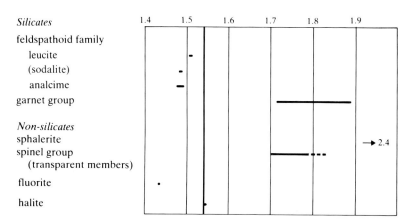

Appendix B 2*V* size and sign of biaxial minerals

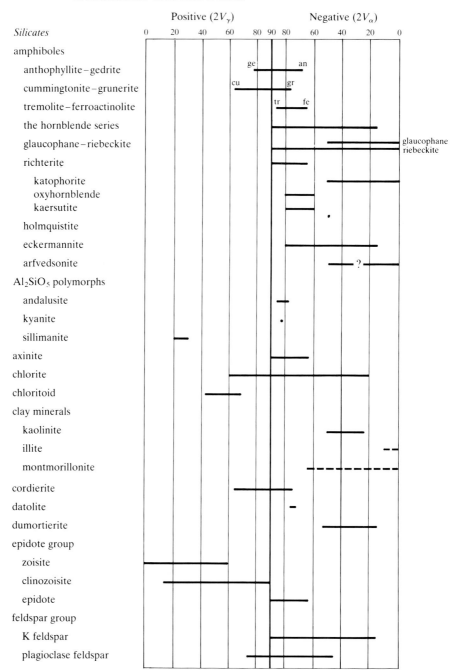

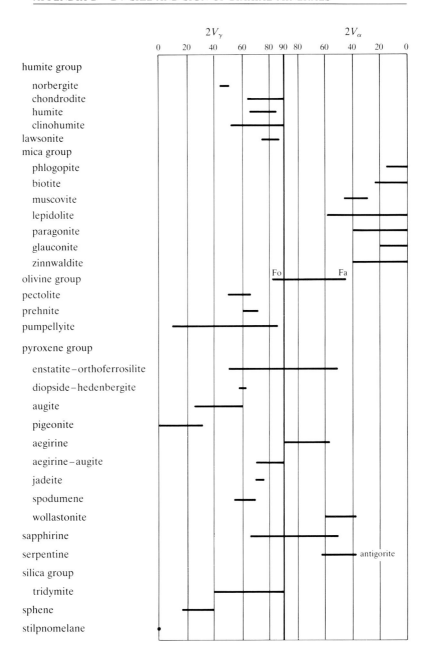

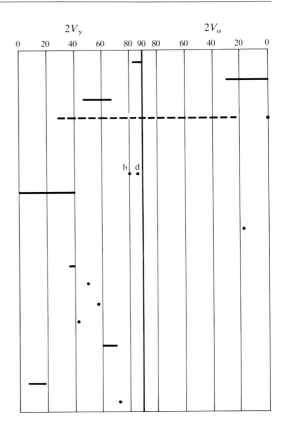

staurolite

talc

topaz

zeolite

Non-silicates

diaspore – boehmite

gibbsite

carbonates

  aragonite

sulphates

  barite

  celestite

  gypsum

  anhydrite

wolframite

phosphates

  monazite

  wavellite

Appendix C Properties of ore minerals

Mineral	Formula	R (%)	Colour	VHN	Anisotropy	Distinguishing properties/resemblance (associations)	See page
acanthite	Ag₂S low-temperature polymorph	30–31	light grey (greenish)	20–60	distinct	twin lamellae (Pb–Sb–As–Bi–Ag–Au)	—
alabandite	MnS	23	light grey	240–250	isotropic	lamellar twinning; brown or green internal reflections (sulphides, Mn-carbonates)	—
anatase	TiO₂ low-temperature polymorph	19	light grey	600–700	weak	abundant colourful internal reflections; resembles rutile (Fe–Ti–O, pyrite)	183
argentite	Ag₂S high-temperature polymorph	30	light grey (greenish)	20–60	isotropic	(Ag-sulphides, Cu–Pb–S, Au)	—
armalcolite	(Fe,Mg)Ti₂O₅	13–14	grey (brownish)	?	moderate	barrel-shaped (Fe–Ti–O, moon)	—
arsenopyrite	FeAsS	52	white	1050–1130	distinct	rhomb-shaped, twinning, zoning (sulphides, oxides, Au–Bi–Te–Sn, W)	197
bismuth	Bi	60–67	bright white	10–20	distinct	tarnishes brown, multiple twinning (Co–Ni–As–S, Au–Bi, Te)	—

bismuthinite	Bi$_2$S$_3$	37–48	white (bluish)	70–220	very strong	fibrous, straight extinction (Co–Ni–As–S, Au–Bi–Te, Mo, Sn, W)	—
blaubleibender covellite	Cu$_{1+x}$S	↑	↑	↑	↑	resembles covellite, but R_i, blue in oil (Cu–Fe–S)	201
bornite	Cu$_5$FeS$_4$	22	light pinkish brown	100	very weak	tarnishes to blue or purple; intergrowths with chalcopyrite (Cu–Fe–S)	198
boulangerite	Pb$_5$Sb$_4$S$_{11}$	37–41	light grey (greenish)	90–180	distinct	stronger anisotropy than bournonite (Pb–Sb–S, Cu–Fe–S)	—
bournonite	CuPbSbS$_3$	34–36	light grey (bluish)	160–210	weak	common twinning (Pb–Sb–S, Cu–Fe–S)	—
braunite	Mn$_7$SiO$_{12}$	19–20	light grey (brownish)	920–1200	weak	resembles some manganese oxides; may be magnetic (Mn–Fe–Si–O, mainly in metamorphic rocks)	—
bravoite	(Fe,Ni,Co)S$_2$	31 → 54 Ni,Co→Fe	light grey to white (brownish)	670–1540	isotropic	colour zoning (pyrite and other sulphides)	208
carrollite	Co$_2$CuS$_4$	43	white (pinkish)	500–590	isotropic	resembles linnaeite (Co–Cu–Fe–S)	—
cassiterite	SnO$_2$	11–12	grey	1240–1470	distinct	common twinning; strong colourless to brown internal reflections; cleavage (W, Bi, As, B, sulphides)	176
chalcocite	Cu$_2$S	33	light grey (bluish)	80–90	weak (colourful)	lancet-shaped, twinning (Cu–Fe–S)	198

Mineral	Formula	R (%)	Colour	VHN	Anisotropy	Distinguishing properties/resemblance (associations)	See page
chalcopyrite	$CuFeS_2$	44–46	yellow	180–200	weak	twinning; more yellow and softer than pyrite; as inclusions in sphalerite (Cu–Fe–Ni–S, sulphides)	199
chromite	$FeCr_2O_4$	13	grey (brownish)	1270–1460	isotropic (weak anisotropy)	rounded octahedra; resembles magnetite but non-magnetic (Fe–Ti–O)	177
cinnabar	HgS	24–29	light grey (bluish)	80–160	moderate	multiple twinning; abundant red internal reflections; rare (Hg–Sb–S, Fe–S)	200
cobaltite	$CoAsS$ orthorhombic	51	white (pinkish)	1100–1350	weak	often idiomorphic cubic; colour zonation; cleavage traces (Cu–Fe–S, Co–Ni–As–S)	201
cohenite	Fe_3C		white	—	weak	resembles iron (Fe, Fe–O, Fe–Ni–S, meteorites)	—
copper	Cu	75	metallic pink (tarnishes)	80–100	isotropic	scratches easily (Cu–O, Cu–Fe–S)	169
covellite	CuS	7–24	blue to bluish light grey	50–140	very strong (fiery orange)	plates and flakes; pleochroic (Cu–Fe–S)	201
cryptomelane	$\sim K_2Mn_8O_{16}$	27	light grey	530–1050	distinct	fibrous, botryoidal; resembles psilomelane; straight extinction (Fe–Mn–O)	—

Mineral	Formula	R	Colour	Anisotropy	Remarks	Page
cubanite	$CuFe_2S_3$ orthorhombic	35–39	light grey (yellowish brown)	strong	lamellae within pyrrhotite, chalcopyrite (Cu–Fe–S)	—
cubanite	$CuFe_2S_3$ cubic	35	light grey (pinkish)	isotropic	intergrown with orthorhombic cubanite (Cu–Fe–S)	—
cuprite	Cu_2O	26	light grey (bluish)	strong	deep red internal reflections (Cu, Fe-OH, Cu–Fe–S, Ag)	—
digenite	Cu_9S_5	21	light grey (bluish)	isotropic	(Cu–Fe–S)	198
djurleite	$Cu_{1.96}S$	31–32	light grey (bluish)	weak	resembles chalcocite	199
electrum	(Au,Ag)	85	light yellow	isotropic	resembles gold (Au–Te–Bi–Cu–Fe–As–Sb–Pb–S)	170
enargite	Cu_3AsS_4	24–26	light grey (pinkish)	strong colourful	cleavage $\parallel$ (110) (Cu–Fe–Sb–As–S)	—
galena	PbS	43	white	isotropic	triangular cleavage pits (Pb–Ag–Sb–As–S, sulphides)	202
gersdorffite	$(Ni,Co,Fe)AsS$	46–57	white (pinkish)	isotropic	zoning, cleavage $\parallel$ (100) gives triangular pits (Fe–Co–Ni–As–S)	—
glaucodot	$(Co,Fe)AsS$ orthorhombic	45–50	white	distinct	idiomorphic, cleavage; as inclusions in cobaltite (Co–Ni–As–S)	—
goethite	α-$FeOOH$ orthorhombic	16–18	grey	distinct	colloform, botryoidal or elongate crystals; red to brown internal reflections (in limonite, Fe-minerals, gossans)	167

Mineral	Formula	R (%)	Colour	VHN	Anisotropy	Distinguishing properties/resemblance (associations)	See page
gold	Au	76	bright yellow	30–35	isotropic	very bright; as inclusions in sulphides; in fractures; soft (Au–Te–Bi: Cu–Fe–As–Sb–Pb–S)	170
goldfieldite	$Cu_3(Te,Sb)S_4$	30	light grey (brownish)	290–340	isotropic	zoned (Fe–Zn–S. Au–Ag–Te, Bi)	—
graphite	C	6–26	dark grey (brownish) to grey	10	strong	deformed flakes; bireflectance strong; cleavage (graphite schists, graphite "veins")	170
haematite	Fe_2O_3	26–30	light grey	1000–1100	strong	tabular crystals, microcrystalline masses; lamellar twinning; weak bireflectance (Fe–Ti–O)	179
hydrocarbon		< 5	dark grey		isotropic	rounded grains, interstitial masses, frosted surface, low reflectance but no internal reflections (sedimentary rocks. barite, carbonate, sulphide veins, U)	—
ilmenite	$FeTiO_3$	17–20	light grey (slightly pinkish)	560–700	moderate	occasional twinning; lamellar inclusions of hematite (Fe–Ti–O)	180
iron	Fe	58	bright white	160	isotropic	rounded grains (Fe–Ni–S. moon, meteorites)	—

Name	Formula		Colour			Remarks	
jacobsite	$(Mn,Fe,Mg)(Fe,Mn)_2O_4$	19	grey (brownish–greenish)	660–710	isotropic	rounded grains; fine aggregates; strongly magnetic; resembles braunite (Mn-minerals; Fe–OH; in metamorphic rocks)	—
jamesonite	$Pb_4FeSb_6S_{14}$	36–44	light grey (greenish)	70–90	strong	acicular with cleavage and twin lamellae parallel to length (Fe–Pb–Sb–Ag–S)	—
kamacite	(Fe,Ni)	60	white (bluish)		isotropic	(Fe–Ni–S, Fe–Ti–Cr–O, meteorites)	—
lepidocrocite	γ-FeOOH	11–18	grey	400	very strong	red to brown internal reflections (in limonite, Fe-minerals, gossans)	167
limonite (see goethite and lepidocrocite)	$FeO.OH.nH_2O$	16–19	bluish grey	690–820	strong (colourful)	abundant brown to red internal reflections (replaces iron minerals)	167
linnaeite	Co_3S_4	45–50	white (pinkish)	350–570	isotropic	cleavage $\parallel$ (100) (Cu–Fe–Ni–S)	—
livingstonite	$HgSb_4S_8$	31–40	light grey	90–130	strong (colourful)	scarce red internal reflections; more opaque than cinnabar (Hg–Sb–S)	—
loellingite	$FeAs_2$	53–55	white (yellowish)	860–920	very strong (colourful)	common twinning (Fe–Ni–As–S, Cu–Fe–S, U, Sn)	—
mackinawite	$(Fe,Ni,Co,Cu)S$	16–41	light grey (pink red)	50–60	very strong	lamellae in Cu–Fe–S or Fe–Ni–S phases (Cu–Fe–Ni–S)	—
maghemite	γ-Fe_2O_3	24	light grey (bluish)	410	isotropic	magnetic (Fe–Ti–O)	—

Mineral	Formula	R (%)	Colour	VHN	Anisotropy	Distinguishing properties/resemblance (associations)	See page
magnetite	Fe_3O_4	21	light grey (often pinkish)	500–790	isotropic	magnetic; octahedra, rounded; lamellar inclusions of hematite or ilmenite ($Fe–Ti–O$)	182
manganite	MnOOH	14–20	grey (brownish)	630–740	strong	elongate crystal aggregates; cleavages; twin lamellae; red internal reflections; alters to pyrolusite ($Fe–Mn–O–Si$, veins)	—
marcasite	FeS_2 orthorhombic	49–55	white (slightly yellowish)	910–1000	strong (colourful)	radiating aggregates of twins; intergrown with pyrite (sulphides in sedimentary rocks, veins)	203
melnicovite–pyrite	$\sim FeS_2$	50	white (brownish)		isotropic or weak anisotropy	very fine grained aggregate; minute pits; banding (sulphide in sedimentary rocks)	208
miargyrite	$AgSbS_2$	31–34	light grey (bluish)	90–120	strong	twinning; scarce red internal reflections; as inclusions in galena ($Ag–Pb–Sb–S$)	—
millerite	NiS	50–56	yellow	190–380	strong	aggregates of acicular grains; common twinning ($Cu–Fe–Ni–S$)	—
molybdenite	MoS_2	19–39	grey to white	20–30	very strong	flakes, platelets, hexagonal; poor polish ($Bi–Te–Au$, Sn, W, sulphides)	204

Mineral	Formula	R%	Colour	VHN	Anisotropy	Remarks	
niccolite	NiAs	48–52	white (orange or pinkish)	330–460	very strong	cataclased grains in radiating botryoidal masses; twinning (Ni–Co–Ag–As–U, sulphides)	155
orpiment	As$_2$S$_3$	23–27	light grey	20–50	strong	radiating aggregates; abundant strong white to yellow internal reflections (realgar)	—
pentlandite	(Fe,Ni)$_9$S$_8$	47	white (slightly yellowish)	270–290	isotropic	triangular cleavage pits; alteration along octahedral parting; as lamellae in pyrrhotite (Cu–Ni–Fe–S)	205
perovskite	(Ca,Na,A)(Ti,Nb)O$_3$ A = rare earths	16	grey	920–1130	very weak	cubic octahedral habit; lamellar twinning; strong white to brown internal reflections (Cu–Fe–Ti–O, alkaline igneous rocks)	—
pitchblende	UO$_{2.3}$	16	grey	670–800	isotropic	radioactive; botryoidal masses, variation in R; shrinkage cracks (Ni–Co–Ag–Bi; Au, sulphides)	185
platinum	Pt	70	bright white (yellowish)	290–340	isotropic (incomplete extinction)	zoning; exsolved, intergrown phases of Pt group elements (Pt–Ir–Os–Rh–Ru–Pd, Fe–Cr–O, Cu–Fe–S)	—
proustite	Ag$_3$AsS$_3$	25–28	light grey (bluish)	70–110	strong	distinct bireflectance; red internal reflections (Ag–Pb–Sb–As–S)	205
pseudobrookite	Fe$_2$TiO$_5$	15	grey		distinct	red to yellow internal reflections; resembles rutile (Fe–Ti–O)	—

Mineral	Formula	R (%)	Colour	VHN	Anisotropy	Distinguishing properties/resemblance (associations)	See page
psilomelane	$(Ba,Mn,Al,Fe,Si)_3$ $Mn_8O_{16}\cdot(O,OH)_6$	$15\rightarrow30$	grey →light grey (bluish)	200–810	strong	fine crystalline aggregates, botryoidal; resembles cryptomelane (Mn–O, Fe–OH: a weathering product)	—
pyrargyrite	Ag_3SbS_3	28–30	light grey (bluish)	50–130	strong	distinct bireflectance: red internal reflections (Ag–Pb–Sb–As–S)	205
pyrite	FeS_2 cubic	54	white (yellowish)	1000–2000	isotropic (weak anisotropy)	idiomorphic, fractured; framboids (common sulphide in all rock types)	206
pyrolusite	MnO_2	30–36	light grey (yellowish)	80–1500	very strong if well crystallized	coarse to cryptocrystalline; cleavage twinning (Fe–Mn–O–OH; a weathering product; veins)	—
pyrrhotite	$Fe_{1-x}S$	35–40	white (brownish or pinkish)	370–410	strong (colourful)	polycrystalline aggregates, twinning; alters readily; magnetic (Cu–Fe–Ni–S, Fe–Ti–O)	209
rammelsbergite	$(Ni,Co,Fe)As_2$	57–61	bright white	630–760	strong (colourful)	radiating zoned aggregates, skeletal; common lamellar twinning (Co–Ni–Fe–As, Ag, Ag–Sb–S)	—
realgar	AsS	20–21	light grey	50–60	strong	abundant strong yellow to red internal reflections; interstitial (orpiment, Fe–As–Sb–S, Fe–OH)	—
rutile	TiO_2 high-temperature polymorph	20–23	light grey (bluish)	890–970	strong	abundant bright internal reflections; twinning; acicular (Fe–Ti–O, Fe–S)	183

safflorite	(Co,Fe,Ni)As$_2$	55(–66)	white (bluish)	790–880	strong (zoned)	concentric radiating aggregates with other minerals; star shaped twins (Co–Ni–Fe–As, Bi, U, Ag, sulphides)	—
scheelite	CaWO$_4$	10	dark grey	380–460	distinct	rhombohedral cleavages; abundant white internal reflections (wolframite, Cu–Fe–As–S, Bi–S, Sn, Au)	—
schriebersite	(Fe,Ni)$_3$P		white		weak	elongate inclusions in iron minerals (Fe–Ni–S, meteorites, moon)	—
silver	Ag	95	bright white (metallic)	50–120	isotropic	scratches easily; skeletal or dendritic (Co–Ni–Fe–As, oxidized sulphides)	171
skutterudite	(Co,Ni,Fe)As$_{3-x}$	54 variable	white (yellowish bluish or pinkish)	600–920	isotropic (weak anisotropy)	cleavage traces, compositional zoning (Co–Ni–F–As–S, Au, Ag, U, Mo, Bi)	—
sphalerite	(Zn,Fe)S	17	grey	200–220	isotropic	colourless to red internal reflections; irregular fractures and cleavage pits (sulphides)	209
spinel	MgAl$_2$O$_4$	8	dark grey	860–1650	isotropic	octahedral, rounded; internal reflections; as inclusions in magnetite (Fe–Ti–O)	184
stannite	Cu$_2$SnFeS$_4$	26–28	light grey (greenish to brownish)	210–270	strong	cleavage, triangular pits, pleochroism, lamellar and cross hatch twinning (sulphides, Sn–W–As–Bi–Au)	—
stibnite	Sb$_2$S$_3$	30–47	light grey to white (slightly brownish)	70–90	very strong	acicular or bladed; distinct bireflectance; cleavage traces; deformation twinning (Ag–Pb–Sb–S, Fe–S, Au, Hg–S)	211

Mineral	Formula	R (%)	Colour	VHN	Anisotropy	Distinguishing properties/resemblance (associations)	See page
taenite	(Fe,Ni)		white (yellowish)		isotropic	lamellae in kamacite (Fe–Ni–S, Fe–Ti–Cr–O, meteorites)	—
tennantite	$Cu_{10}(Zn,Fe)_2(As,Sb)_4S_{13}$	30	light grey	290–390	isotropic	polycrystalline aggregates (Cu-Fe-S, galena, sphalerite)	212
tetradymite	Bi_2Te_2S	55–60	white (yellowish)	30–50	distinct	hexagonal cross sections, basal cleavage (Au–Bi–Te–S, Cu–Fe–S)	—
tetrahedrite	$Cu,Ag)_{10}(Zn,Fe,Hg)_2(Sb,As)_4S_{13}$	32	light grey	240–370	isotropic	polycrystalline aggregates (Cu–Fe–S, galena, sphalerite)	212
titanomagnetite	$(Fe,Ti)_3O_4$	17	grey (brownish or pinkish)	720–730	isotropic (weak anisotropy)	homogeneous only if formed by rapid cooling, otherwise intergrown with Fe–Ti–O phases (Fe–Ti–O in igneous and metamorphic rocks)	182
todorokite	$(H_2O)_{<2}Mn_{<8}(O,OH)_{16}$	20	light grey	—	strong	columnar to fibrous aggregates, botryoidal; cleavage traces (Mn–Fe–O; deep-sea Mn nodules)	—
troilite	FeS		yellow (brownish)		strong (colourful)	resembles pyrrhotite; rare (Fe–Ni–S, moon, meteorites)	—
ulvospinel	Fe_2TiO_4	17	brownish grey		isotropic	as fine intergrowths in titaniferous magnetite; octahedra (Fe–Ti–Cr–O, igneous, moon)	182
uraninite	UO_2	17	grey	780–840	isotropic	associated with pitchblende (Ni–Co–Ag–Bi, Au)	185
wolframite	$(Fe,Mn)WO_4$	15–16	grey (slightly brownish)	320–390	moderate	bladed crystals, simple twins, reddish brown internal reflections (Sn, Au, Bi)	213
würtzite	ZnS hexagonal	↑	↑	↑		resembles sphalerite, but rare	—

Appendix D Mineral identification chart

This simple chart shows the optical properties of the common ore minerals listed in order of relative polishing hardness, and it can be used as an aid to mineral identification. Reflectance values ($R\%$) are given with minerals plotted in their correct position, but some minerals are plotted in brackets in a second position because of their variable appearance.

Procedure

(1) Determine whether the unknown mineral is isotropic, weakly anisotropic or distinctly anisotropic. Weak anisotropy is seen using slightly uncrossed polars, whereas distinct anisotropy is easily visible with exactly crossed polars.

(2) Note whether the mineral is colourless, slightly coloured or coloured in PPL; white to grey minerals are considered to be colourless for this purpose. Take care to consider the colour in relation to several adjoining minerals.

(3) Estimate the brightness (reflectance) of the mineral as a percentage. This is usually rather difficult unless some minerals in the section have already been identified. An uncertainty of ± 5 is typical for estimates of $R\%$ unless a good reference mineral is also in the section. If the mineral is distinctly bireflecting then estimate the minimum and maximum reflectance values.

(4) Whether the mineral is hard or soft can usually be determined by its polishing behaviour, e.g. pits persist in hard minerals, whereas soft minerals scratch easily. Also, hard minerals tend to stand proud of the surface, whereas soft minerals are scoured out. The Kalb light line can be used to compare polishing hardness with other minerals in the section.

If the properties of the unknown mineral appear to correspond to one of the minerals on the chart, then it is best to check all the mineral's characteristics with the information given for the selected mineral in the description of the ore minerals before concluding the identification.

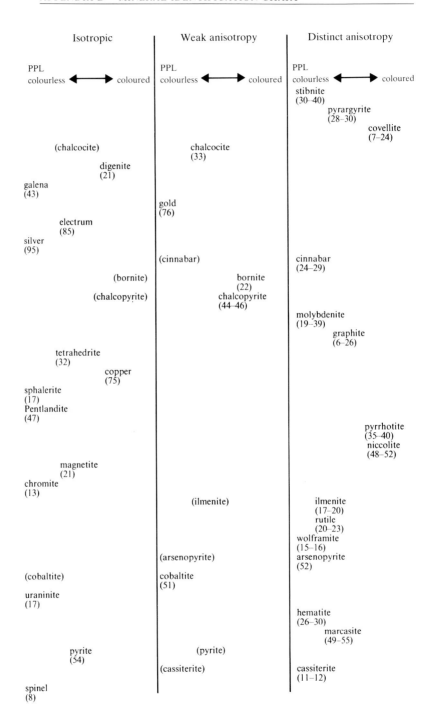

Isotropic	Weak anisotropy	Distinct anisotropy
PPL colourless ⟷ coloured	PPL colourless ⟷ coloured	PPL colourless ⟷ coloured

stibnite (30–40)

pyrargyrite (28–30)

covellite (7–24)

(chalcocite)

chalcocite (33)

digenite (21)

galena (43)

gold (76)

electrum (85)

silver (95)

(cinnabar)

cinnabar (24–29)

(bornite)

bornite (22)

(chalcopyrite)

chalcopyrite (44–46)

molybdenite (19–39)

graphite (6–26)

tetrahedrite (32)

copper (75)

sphalerite (17)

Pentlandite (47)

pyrrhotite (35–40)

niccolite (48–52)

magnetite (21)

chromite (13)

(ilmenite)

ilmenite (17–20)

rutile (20–23)

wolframite (15–16)

(arsenopyrite)

arsenopyrite (52)

(cobaltite)

cobaltite (51)

uraninite (17)

hematite (26–30)

marcasite (49–55)

pyrite (54)

(pyrite)

(cassiterite)

cassiterite (11–12)

spinel (8)

290

Appendix E   Gangue minerals

The gangue minerals referred to here are the minerals that commonly accompany ore minerals in hydrothermal deposits. Although they are transparent, and are best studied using transmitted-light microscopy, it is useful to be able to *recognize* the common gangue minerals in reflected light (see Fig. 1.8). The minerals listed all have low reflectance values, but the eye can determine small differences in brightness even at these low values. The carbonates are exceptional in having large birefringences and this results in distinct bireflectance: the resulting strong anisotropy is often masked by internal reflections. It is relatively easy to recognize a mineral as being a carbonate. As well as using optical properties, physical and textural properties can be used in recognizing the gangue minerals:

Quartz   Lack of cleavage but irregular fractures; crystal shape, especially pyramidal terminations; lack of alteration.

Barite   Several sets of cleavage traces; bladed or tabular crystals; radiating aggregates.

Fluorite   Octahedral cleavage giving up to three cleavage traces and triangular cleavage pits; cubic crystals.

K-feldspar Several cleavage traces; alteration.

Carbonates Rhombohedral cleavage resulting in up to three cleavage traces; multiple twinning; rhomb-shaped crystals.

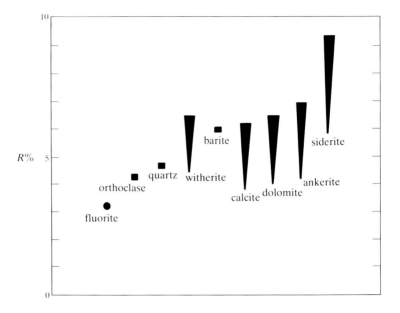

Note that carbonates have a pronounced bireflectance and that R_o, which is shown by *all* grains of a carbonate mineral, is indicated by the wide end (maximum value) of the reflectance range.

Mineral	Refractive indices		Reflectances	
quartz SiO_2	$n_o = 1.544$	$n_e = 1.553$	$R_o = 4.6$	$R_e = 4.7$
barite $BaSO_4$	$n_\alpha = 1.636$	$n_\gamma = 1.648$	$R_\alpha = 5.8$	$R_\gamma = 6.0$
fluorite CaF_2	$n = 1.434$		$R = 3.2$	
orthoclase $KAlSi_3O_8$	$n_\alpha = 1.518$	$n_\gamma = 1.522$	$R_\alpha = 4.2$	$R_\gamma = 4.3$
calcite $CaCO_3$	$n = 1.658$	$n_e = 1.486$	$R_o = 6.1$	$R_e = 3.8$
dolomite $CaMg(CO_3)_2$	$n_o = 1.679$	$n_e = 1.500$	$R_o = 6.4$	$R_e = 4.0$
ankerite $Ca(Fe,Mg)(CO_3)_2$	$n_o = 1.710$	$n_e = 1.515$	$R_o = 6.9$	$R_e = 4.2$
siderite $FeCO_3$	$n_o = 1.875$	$n_e = 1.635$	$R_o = 9.3$	$R_e = 5.8$
witherite $BaCO_3$	$n_\alpha = 1.529$	$n_\gamma = 1.677$	$R_\alpha = 4.4$	$R_\gamma = 6.4$

Bibliography

Atkin, B. P. & P. K. Harvey 1982. NISOMI-81: an automated system for opaque mineral identification in polished section. *Process mineralogy II: applications in metallurgy, ceramics and geology*, 77–91. Conference Proceedings, The Metallurgical Society of AIME, Dallas.

Bastin, E. S. (1950). *Interpretation of ore textures*. Memoirs of the Geological Society of America, 45.

Bloss, F. D. 1971. *Crystallography and crystal chemistry*. New York: Holt, Rinehart and Winston.

Bowen, N. L. & J. F. Schairer 1935. The system $MgO–FeO–SiO_2$. *American Journal of Science* **229**, 151–217.

Bowie, S. H. U. & P. R. Simpson 1977. Microscopy: reflected-light. In *Physical methods in determinative mineralogy*, 2nd edn, J. Zussman (ed.). London: Academic Press.

Bowie, S. H. U. & P. R. Simpson 1980. *The Bowie–Simpson System for the microscopic determination of ore minerals. First Students' Issue*. London: McCrone Research Associates.

Bowie, S. H. U. & K. Taylor 1958. A system of ore mineral identification. *Mining Magazine (London)* **99**, 265–77, 337–45.

Buddington, A. F. & D. H. Lindsley 1964. Iron–titanium oxide minerals and synthetic equivalents. *Journal of Petrology* **5**, 310–57.

Cameron, E. N. 1961. *Ore microscopy*. New York: John Wiley.

Cervelle, B. 1979. The reflectance of absorbing anisotropic minerals. In *Proceedings of the 1974 ore microscopy summer school at Athlone*, M. J. Oppenheim (ed.), 49–57. Special Paper no. 3. Geological Survey of Ireland.

Commission of Ore Microscopy 1977. *IMA/COM quantitative data file*, N. F. M. Henry (ed.). London: Applied Mineralogy Group, Mineralogical Society.

Craig, J. R. & D. J. Vaughan 1981. *Ore microscopy and ore petrography*. New York: John Wiley.

Criddle, A. J. & C. J. Stanley (eds) 1986. *The quantitative data file for ore minerals of the Commission on Ore Microscopy of the International Mineralogical Association*, 2nd issue. London: British Museum (Natural History).

Deer, W. A., R. A. Howie & J. Zussman 1962. *Rock-forming minerals*, Vols 1–5. London: Longman.

Deer, W. A., R. A. Howie & J. Zussman 1966. *An introduction to rock-forming minerals*. London: Longman.

Deer, W. A., R. A. Howie & J. Zussman 1978. *Rock-forming minerals: single chain silicates*, vol. 2A. London: Longman.

Edwards, A. B. 1947. *Textures of the ore minerals*. Melbourne: Australasian Institute of Mining and Metallurgy.

Embrey, P. G. & A. J. Criddle 1978. Error problems in the two media method of deriving the optical constants n and k from measured reflectances. *American Mineralogist* **63**, 853–62.

Freund, H. (ed.) 1966. *Applied ore microscopy*. New York: Macmillan.

Galopin, R. & N. F. M. Henry 1972. *Microscopic study of opaque minerals*. London: McCrone Research Associates.

Hallimond, A. F. 1970. *The polarising microscope*. York: Vickers Instruments.

Henry, N. F. M. (ed.) 1970–8. *Mineralogy and Materials News Bulletin for Microscopic Methods*. London: Mineralogical Society.

Holdaway, M. H. 1971. Stability of andalusite and the aluminium silicate phase diagram. *American Journal of Science* **271**, 97–131.

IMA/COM 1977. *Quantitative data file of the Commission on Ore Microscopy of the International Mineralogical Association*, 1st issue. London: McCrone Research Associates.

Judd, D. B. 1952. *Colour in business, sciences and industry*. New York: John Wiley.

Kerr, P. F. 1977. *Optical mineralogy*. New York: McGraw-Hill.

Leake, B. E. 1978. Nomenclature of amphiboles. *Canadian Mineralogist* **16**, 501–20.

Lister, B. 1978. *Ore polishing*. Institute of Geological Sciences Report 78/27. London: HMSO.

Nickless, G. (ed.) 1968. *Inorganic sulfur chemistry*. Amsterdam: Elsevier.

Oppenheim, M. J. (ed.) 1979. *Proceedings of the 1974 ore microscopy summer school at Athlone*. Special Paper no. 3. Geological Survey of Ireland.

Palache, C., H. Berman & C. Frondel (eds) 1962. *Dana's system of mineralogy*, 7th edn, vols I, II & III. New York: John Wiley.

Pauling, L. & E. H. Neuman 1934. The crystal structure of binnite, $(Cu,Fe)_{12}As_4S_{13}$ and the chemical composition and structure of minerals of the tetrahedrite group. *Zeitschrift für Kristallographie* **88**, 54–62.

Phillips, W. R. & D. T. Griffin 1981. *Optical mineralogy: the non-opaque minerals*. New York: W. H. Freeman.

Picot, P. & Z. Johan 1977. *Atlas des minéraux métalliques*. Mémoires du Bureau de Recherches Géologiques et Minérals, Orléans, no. 90.

Picot, P. & Z. Johan 1982. *Atlas of ore minerals*. Amsterdam: Elsevier.

Piller, H. 1966. Colour measurements in ore microscopy. *Mineralium Deposita* **1**, 175–92.

Piller, H. 1977. *Microscope photometry*. Berlin: Springer.

Ramdohr, P. 1969. *The ore minerals and their intergrowths*. Oxford: Pergamon.

Ramdohr, P. 1980. *The ore minerals and their intergrowths*, 2nd edn. Oxford: Pergamon.

Ribbe, P. H. 1974. *Short course notes*. Vol. 1, *Sulfide mineralogy*. Mineralogical Society of America.

Rumble, D. III (ed.) 1976. *Short course notes*. Vol. 3, *Oxide minerals*. Mineralogical Society of America.

Shepherd, T., A. H. Rankin & D. H. M. Alderton 1985. *A practical guide to fluid inclusion studies*. Glasgow: Blackie.

Shuey, R. T. 1975. *Semiconducting ore minerals*. Amsterdam: Elsevier.

Smith, J. V. 1974. *Feldspar minerals*, vols 1–3. Berlin: Springer.

Tuttle, O. F. & N. L. Bowen 1958. Origin of granite in the light of experimental studies in the system $NaAlSi_3O_8$–$KAlSi_3O_8$–SiO_2–H_2O. Memoirs of the Geological Society of America, 74.

Uytenbogaardt, W. & E. A. J. Burke 1971. *Tables for microscopic identification of ore minerals*. Amsterdam: Elsevier.

Vaughan, D. J. & J. R. Craig 1978. *Mineral chemistry of metal sulfides*. Cambridge: Cambridge University Press.

Wahlstrom, E. E. 1959. *Optical Mineralogy*, 2nd edn. New York: John Wiley.

Index

Page numbers in **bold** refer to principal entries

Eutaxitic > banded structure